The American Fisherman

The American Fisherman

How Our Nation's Anglers Founded, Fed, Financed, and Forever Shaped the U.S.A.

Willie Robertson
and
William Doyle

 For information, address HarperCollins Publishers, 195 Broadway, New York, NY 10007.

HarperCollins books may be purchased for educational, business, or sales promotional use. For information, please e-mail the Special Markets Department at SPsales@harpercollins.com.

FIRST HARPERLUXE EDITION

ISBN: 978-0-06-249692-8

HarperLuxe™ is a trademark of HarperCollins Publishers.

Library of Congress Cataloging-in-Publication Data is available upon request.

16 17 18 19 20 ID/RRD 10 9 8 7 6 5 4 3 2 1

To the

AMERICAN FISHERMAN,

past, present and future.

And to everyone

who loves fish.

Contents

An American Fisherman (me, pre-beard) with a good day's catch of king salmon on the Sacramento River, California. (Robertson Family Collection)

Introduction

And God said, Let us make man in our image, after our likeness: and let them have dominion over the fish of the sea, and over the fowl of the air, and over the cattle, and over all the earth, and over every creeping thing that creepeth upon the earth.

—Genesis 1:26, King James Version

Perhaps I should not have been a fisherman, he thought. But that was the thing that I was born for.

—Ernest Hemingway, *The Old Man and the Sea*

All Americans believe that they are born fishermen.

—John Steinbeck

From the day America was born, fishing has been a part of the American soul.

For Americans, fishing is much more than a sport.

It is a lifestyle of adventure, escape, and contemplation, of peace and spirituality, and bonding with nature, friends, and family. It can be a hobby, a passion, and a full-time job.

It is a way to challenge the soul, test your limits, and put food on the family table. It's how parents and children discover what they're made of together.

The Biblical faith that inspired America is often symbolized by fish, as symbols of life, sustenance, and redemption, including Jesus Christ's recruitment of fishermen as his first disciples, the story of the loaves and the fishes, and his post-resurrection appearance to Simon Peter and the fishermen at the Sea of Galilee.

George Washington was a commercial fisherman. Native Americans and European settlers were nourished on countless fish from America's lakes, streams, and oceans. Washington and Lincoln's troops were fortified at key moments by fish. A boatload of presidents loved fishing, including Chester Arthur, Grover Cleveland, Calvin Coolidge, Herbert Hoover, Franklin Delano Roosevelt, Dwight Eisenhower, Jimmy Carter, and both Bushes.

Great states like Massachusetts, New York, Florida, Maryland, Louisiana, Texas, California, Oregon, Washington, and Alaska were built in large part upon commercial fishing. After World War II, sportfishing rose to become a multibillion-dollar passion for tens of millions of Americans. Fishing has inspired great American literature, painting, and poetry.

And fishing has directly shaped the creation, life, and growth of the United States of America.

Today, fishing is America's number one pastime. There are more than 33 million fishermen and women in America. That's more than the number of people who play golf and tennis combined! Our numbers are growing fast—by 11 percent in the last five-year federal survey. We spend $48 billion a year on fishing gear, licenses, trips, and events. We contribute over $1 billion a year to protect the environment. In fact, American fishermen, along with hunters, are the world's most powerful conservationists, contributing billions of dollars to keep American lakes, rivers, streams, and oceans pure and safe for aquatic wildlife. We enjoy all kinds of fishing in fresh and salt water, like spincasting, spinning, baitcasting, and fly-fishing. Some of us even like to hunt underwater with a speargun, and others prefer to ice-fish through a hole in a frozen lake.

I can think of three things that help make America great—God, guts, and gills. The American fisherman delivers all three, and is a perfect icon of the freedom, adventure, ingenuity, and devotion that define the United States of America.

So join us on a journey back to our roots, back to the rivers, lakes, and streams of our youth, back to the days when America was young, our horizons were limitless, and we were nourished by the gifts of God and nature.

Climb aboard on this voyage into the heart of the American Fisherman.

Prologue

An American Fishing Family

Some go to church and think about fishing, others go fishing and think about God.

—Tony Blake

In our family, there was no clear line between religion and fly-fishing.

—Norman Maclean, *A River Runs Through It*

The trees, the shrubs, the flowers, the mosses, the grass, the very stones of the brook, have a message for the true fisherman. There is "a sermon in the stones."

—James Hoadley

One day when I was twelve years old, I grabbed my fishing pole and bucket and snuck off from home.

I was in search of a Fisherman's Paradise.

This could be dangerous, I thought. But I've got to do it.

I walked a mile down the road, along a winding creek and into the thick woods that guarded the property of Judge Harrison. He was a distinguished local justice who had a nice piece of land, and my older brother Jase and I had heard tales of a magical pond somewhere in the woods that was home to a staggering number of fish.

The stories about the pond seemed too fantastic to be true. It sounded like a miracle of Nature, a gift from the heavens beyond your wildest dreams of fishing, where you could fish to your heart's content and the fish would just keep on coming.

The trouble was, the pond was on someone else's private property, a judge no less, and I didn't have an invitation. If I got caught, the judge might call the sheriff, and I'd wind up as a condemned trespasser and fish thief. But I was young, and I guess you could say I was a trifle reckless. Nothing could stop me. I had to see that pond. I had to see if it was real. And I had to get my hands on some of those fish.

The sun was burning away the morning mist on a hot, lazy Louisiana summer day, and the sounds of birds, bees, and crickets buzzed all around me as I marched through the trees with my pole and bucket. I came to a tall picket fence, threw my tackle over, and slipped under the fence. I walked down a grassy slope, and there it was. Harrison's Pond. My jaw dropped.

"Oh, my goodness!" I thought. I was blown away. It was so pretty and so secluded. It looked like not another soul had ever been there.

I saw a little bridge over to the far side of the pond, a perfect spot to fish from. Just seconds after I put my line in, a good-sized bream hit my hook and a bass hit my cork. Bingo. Same thing happened again. As fast as I could drop my line down into the water, the fish just kept on striking, over and over.

It was unbelievable. The Miracle of Harrison's Pond was true! My bucket was full in twenty minutes flat. I ran home as fast as I could to tell my brother the news.

"Jase! *I've found the mother lode of fish!*" We ran back to the pond and dropped our lines. Bang, bang, bang, the fish kept striking. The buckets filled to overflowing. We caught probably over a hundred fish just on that first day. We laughed, we hugged each other, we practically cried tears of joy.

Before long, we were going back and forth to that pond so often that we just stashed our fishing gear at a spot in the woods. If we heard dogs barking or a truck coming down the road in the distance when we were fishing, we'd take off, run for cover, and hide. But we never saw anyone else. It was our secret fishing spot.

Then one day, Jase and I went back to get our stuff, and—all our poles and buckets and everything were gone! Judge Harrison had found it! We knew the game was up and we hightailed it out of there. We had caught nearly every fish out of that pond. It was the end of the line for us.

We never heard a word about it, and Judge Harrison became a good family friend. Years later, we shot some scenes for our TV show at Harrison's Pond, and the good judge was just as nice as can be. I'm still not sure if he appreciates just how many fish we swiped from his pond. If so, he's forgiven us our past transgressions.

I wouldn't recommend sneaking on anyone's property, but at twelve years old, it was a lot of good, wholesome summer entertainment for us, without a doubt.

In fact, that first day at Harrison's Pond was the greatest fishing day of my life.

But how sweetly memories of the past come to one who has appreciated and enjoyed it from his boyhood, whose almost

first penny, after he wore jacket and trowsers, bought his first fish-hook; whose first fishing-line was twisted by mother or sister; whose float was the cork of a physic vial, and whose sinkers were cut from the sheet-lead of an old tea-chest! Thus rigged, with what glad anticipations of sport, many a boy has started on some bright Saturday morning, his gourd, or old cow's horn of red worms in one pocket, and a jack-knife in the other, to cut his alder-pole with, and wandered "free and far" by still pool and swift waters, dinnerless—except perhaps a slight meal at a cherry tree, or a handful of berries that grew along his path—and come home at night weary and footsore, but exulting in his string of chubs, minnows, and sunnies, the largest as broad as his three fingers!

—THADDEUS NORRIS, *THE AMERICAN ANGLER'S BOOK*

No man ever steps in the same river twice, for it's not the same river and he's not the same man.

—HERACLITUS

When you are on the river, ocean, or in the woods, you are the closest to the truth you'll ever get.

—JACK LEONARD

I am an American Fisherman.

I fish for food, I fish for fun, and I fish for family. And I fish to honor God.

The Bible specifically instructs mankind not only to enjoy hunting and fishing, but also to replenish the earth. I try to do that, by fishing responsibly and ethically, and honoring the biblical instruction given in Revelation 7:3, "Hurt not the earth, neither the sea, nor the trees."

I come from a family that's famous for duck and deer hunting, but we're fishermen, too. We don't see much difference between the two. Fishing is just hunting in the water, with a different set of tools and tricks.

As far back as I can remember, I was deeply familiar with the joys and frustrations of fishing. Early on, my family scratched out a living (just barely), through commercial fishing. It was a backwoods, country-style family operation, nothing fancy. Today, I do my fishing in my own state-of-the-art Bass Boat, all tricked out with radar, sonar, digital fish finder, depth finder, GPS, and satellite TV. But back then, all we had was my dad, Phil.

Phil is the original Fish Whisperer, the master of all things in the water. He can just summon fish and they come up to the surface. He knows where they are, how they move and think, and how to catch them. One afternoon, when our family was all dressed up and about to go to Wednesday night church service, he said, "Hang on real quick." We watched him pull out twenty-two bass in the space of twelve minutes.

Phil learned how to fish from his own father, and he taught me and my brothers all the basics—how to tie hooks on, how to tie the string on your pole, how to make the most of things when you didn't have many hooks and not much bait. He taught us how to use grass shrimp, worms, and wasps' nests as bait for different fish. He taught us how to drop hooks in the creek, how to fish with seine nets, milk jugs, catfish poles, and trout lines, and how and where to catch the most fish.

Phil taught us how to chop up snakes and pretty much any other animal to use as crawfish bait. "Crawfish will eat literally anything," he explained. "You stand out there in the water long enough, they'll eat you, too." Every night, he had us fill up a freezer with chopped-up creatures to bait the crawfish traps with the next day. The inside of the freezer was a very scary sight. "Just put a hunk of something dead in there, boys," Phil commanded. "You can pick up anything dead on the road and throw it in, too, just as long as it isn't too far gone."

One day, I shot a big fat water snake and threw him in the freezer. I thought I'd put a hole in his head. I had something to eat, came back, and reached into the freezer to chop him up. To my surprise, that giant snake reared up and gave me a dirty look right in my eyes. He was alive and well! I guess I'd just knocked him out with a flesh wound. Luckily he was in that freezer for a while so he wasn't as lively as he normally would have been, and he didn't bite me. I shot him again and chopped him up into twelve pieces, enough for a dozen crawfish traps.

It got scary when you pulled up the crawfish traps, since there was no telling what might be in them. One time we pulled up a trap and there was a big green, slimy river eel in it. My brother Jase and I kept grabbing this eel and trying to hit him with a paddle but he kept sliding away, so I got the clever idea I was just going to bite his head to make him settle down. I'd always watched my dad crush the skulls of ducks with his teeth to make sure they were dead, so I figured this approach ought to work with an eel.

Let me tell you something. Don't ever try to bite an eel. Because that slime gets between your teeth and it takes three days to get it out! I bit as hard as I could but I could not bite through that eel's skin. So we gave him the cleaver treatment. On another occasion, as my friend Paul gaped in horror, I baited fifteen hooks by biting into perch and tearing them apart with my teeth. I don't recommend that, either!

Jase and I held all-day fishing competitions along Cypress Creek, which ran through our property. All we needed was a hook or a cork or a tight line and some bait. Every fish counted toward the total, no matter how small. I'd catch thirty-two, Jase would land forty, then one day I'd catch fifty-two. If we pulled out a catfish it was like winning the lottery! We'd whoop and holler and run up to the duck-call shed and show our dad.

When he was twenty-eight years old, our father, Phil, decided to repent to the Lord and live a clean, Bible-based life. He told my mother, "Miss Kay, you know what I'm going to do? I'm going to fish the Ouachita River and I'm going to catch catfish and buffalo fish and I'm going to sell them. And I'll make enough money to start a duck-call business." He recalled, "In a lot of ways, I was withdrawing from mainstream society. I was trying to drop back about two centuries to become an eighteenth-century frontier man who relied on hunting and fishing for his livelihood."

At the time, lots of folks thought he was crazy. "So let me get this straight," they'd ask him. "What do you do?"

"Well, I'm fishing the Ouachita River and I'm trying to get this new duck-call product on the market."

Me and my father, Phil, otherwise known as the Fish Whisperer, toting up the catch of king salmon on the Sacramento River. (Robertson Family Collection)

"And you have a master's degree in education?"

"That is correct," Phil would say. "I'm a very smart fisherman, because I have the sheepskin to prove it." Folks would walk off scratching their heads.

Miss Kay and I would take the haul of fish down to the local market, where I'd wheel and deal and try to negotiate the best price so our family could squeeze out a few extra dollars to live on. It was a tough way to make money, but it was good business training for me.

Being the future CEO that I was, I decided to diversify by branching out into the worm business. My grandmother had a boat dock on her property and she charged a dollar a pop for fishermen to use it. Every weekend a ton of fishermen would pull up, so I launched a worm-selling operation at the same spot. I set up an old wooden boat on sawhorses.

My dad said that worms love cow manure, and about a mile up the road there was a farm with cows. So wheelbarrow-by-wheelbarrow, I'd get as much cow dung as I could and wheel it down and mix it with dirt and put it in the boat. Then I dug up every earthworm I could find, thousands of them, filled up the boat, and put some thin boards over the top to create the moisture. When the fishermen came I had a little sign out by the mailbox that said "Worms for a Dollar a Can, or 3 Cents Each." I made some good money with that little worm operation!

I will admit my brother Jase is a more talented fisherman, and he claims that when we were boys, I spent most of my time on my butt eating Pop-Tarts while he caught all the fish. I'm not too sure about that, but he swears it's true.

I love nearly every kind of fishing—with live bait, with lures, with a fly, with nets, with a jig with pork rind or craw-worm dressing or Carolina-rigged plastic lizards and six- to eight-inch Texas-rigged worms—you name it. But I'm not a big saltwater fisherman. The problem is I get too seasick. Most of my saltwater trips end up with me puking my guts out.

When my wife-to-be, Korie, and I were first dating about twenty-five years ago, she and her friend Rachel came to meet my dad. The first thing Phil said to them was: "Have you met my boys, Jason Silas and Willie Jess? They'll make good husbands someday. They're good hunters and fishermen."

I thought I'd show Korie how manly I was, so I took her fishing down at the creek. I was eighteen, she was seventeen. Pretty soon we were raking fish in hand over fist, over forty fish altogether. For some reason I reached down and picked up a clump of mud and threw it at Korie, just for fun. She retaliated. The fishing trip turned into an all-out mud fight. We came back all covered in mud, with a bucket full of fish. So I guess my idea worked! I wooed her in with fish.

Come live with me, and be my love,
And we will some new pleasures prove
Of golden sands, and crystal brooks,
With silken lines, and silver hooks.
There will the river whispering run
Warm'd by thy eyes, more than the sun;
And there th' enamour'd fish will stay,
Begging themselves they may betray.
When thou wilt swim in that live bath,
Each fish, which every channel hath,
Will amorously to thee swim,
Gladder to catch thee, than thou him.

—John Donne, "*The Bait*"

One day, Phil took my son John Luke and his new girlfriend out fishing. They were trying to go on a date, but Phil hijacked their plan. He said he wanted to get some insight into the teenage mind, but I think he was trying to keep an eye on them so they didn't have sex. He took them out fishing, and then supervised them in the slimy business of cleaning the catch.

"Cleaning fish is a great first date," Dad said. "It's romantic!"

When I was growing up, we fished and hunted because we wouldn't eat if we didn't. Ever since then, I've found there are few things I like more than fresh catfish guts on toast. Some days I'll grab a fishing pole and some bait crickets and just sit out there on the banks of the river and see what I can catch. Even when I'm traveling, I try to make a point of getting out to the local fishing holes—it's a special way of connecting to a place. And you'd be surprised how close nature can be even in the busiest city. I once did a store appearance in New York City, and afterward we went out fishing for sea perch in the ocean just off New York harbor. You could see the skyline of New York spread out before you as you were fishing. Not long ago, my father and I went salmon fishing in a beautiful stream right outside Modesto, California.

My kids enjoy fishing, too.

One day when my daughter Sadie was nine years old, I took her out to the little pond in the back of the house. I said, "Sadie, we're going to do supper tonight." It was the first time she went fishing. I wanted her to have the kind of childhood fishing experience we had when we were growing up.

I showed her what to do, and together we must have caught thirty-two little bream. Some of them were real tiny, but we took them home and cleaned every one of them no matter how small, gutted them out, and made fish tacos.

I proudly announced to the family, "Sadie provided this meal."

The smile on her face was a sight to behold.

All the rivers run into the sea; yet the sea is not full; unto the place from whence the rivers come, thither they return again.

—Ecclesiastes 1:7

I will cast down my hook:
The first fish which I bring up
In the name of Christ, King of the Elements,
The poor man shall have for his need:
And the King of the Fishers, the brave Peter,
He will after it give me his blessing.

—Gaelic fisherman's prayer

The first men that our Saviour dear
Did choose to wait upon Him here
Blest fishers were, and fish the last

Food was that He on earth did taste;
I therefore strive to follow those,
Whom He to follow Him hath chose.

—THEODORE WALTON

Look at where Jesus went to pick people. He didn't go to the colleges; he got guys off the fishing docks.

—JEFF FOXWORTHY

When I'm fishing, I feel like I'm closer to God.

I've heard that lots of other folks feel that way, too. Maybe it's because fishing gives us a deep immersion in God's creations of nature, and because it helps us escape the pressures of modern life, slow ourselves way down, and appreciate the infinite beauty of it all.

Whether I'm hunting or fishing, spending time in the wilderness often feels like a spiritual meditation, and a prayer of appreciation to God. It is a sport of endless spiritual surprises and sensations—the serenity of a river at dawn, the bracing cold shock of water on your legs, the force of the currents, the eternities of peaceful waiting, the glimpses of fish in bright clear water, and the mystery of their sudden disappearance. "The traditional Christian wilderness experience is an intense

Nothing says "quality father-and-son time" better than a good day's fishing, even if you don't catch a thing. My son John Luke and I discussing fishing strategies. (Robertson Family Collection)

My daughter Bella expertly reels in a tasty little bream for the family supper. We added sauce and lots of extras and whipped up some delicious fish tacos.
(Robertson Family Collection)

Jesus and the miraculous draught of the fishes.
(New York Public Library Digital Collections)

confrontation with God," wrote scholar Susan Power Bratton. She called the wilderness "the stage which brightly illumines God's power," which "demonstrates God's majesty, love, joy in creation, righteousness, transcendence, and omnipotence."

One enthusiastic fisherman, Henry Ward Beecher, a famous clergyman and abolitionist, put it nicely when he wrote in 1855, "One who believes that God made the world, and clearly developed to us his own tastes and thoughts in the making, can not express what feelings those are which speak music through his heart, in solitary communions with Nature. Nature becomes to the soul a perpetual letter from God, freshly written every day and each hour."

Fishing has always been a big part of my family's life. We see it as an extension of our spiritual life, and we love the fact that the Lord's original disciples were commercial fishermen! We always thought that was very cool, because we were always religious and tried to help bring other people to the Lord. Jesus told his disciples, "I'll make you fishers of men," and we felt we were fishing in a sea of people. They don't just jump in your boat; you have to be smart and bring them in. In a sense, fishing was the foundation of the Bible's New Testament, since Jesus told fisherman Peter, "Upon this rock I will build my church."

The New Testament tells of the miracle of the 153 fish, when seven of the disciples—Peter, Nathanael, Thomas, James and John (the sons of Zebedee), and two others—went fishing one night after the Resurrection of Jesus. They caught nothing that night. The next day, when they took the boat out again, Jesus appeared on the shore, but they didn't recognize him at first.

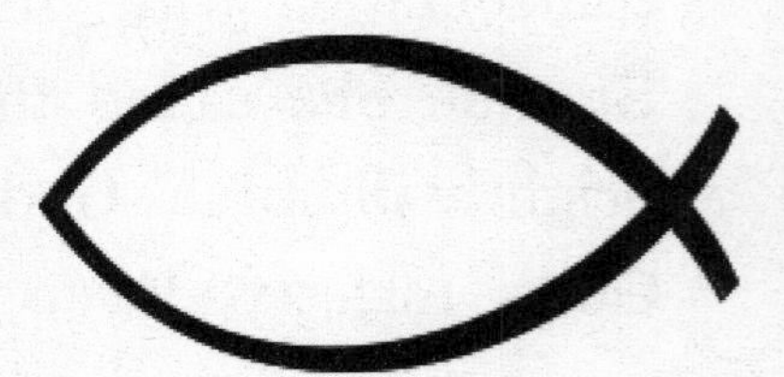

A fish-inspired symbol of Christianity.

"Friends, haven't you any fish?" Jesus called out. They replied no. Jesus replied, "Throw your net on the right side of the boat and you will find some."

When they did this, the fishermen caught so many fish they could barely pull up their net.

Then the stunned disciples realized who the man was and exclaimed joyfully, "It is the Lord!" as Peter jumped into the water to greet him. They all shared a meal of coal-cooked fish and bread.

In Christianity, fish images often symbolize the power and love of Jesus Christ, who fed the multitudes with bread and fish. St. Augustine wrote that "Jesus is a fish that lives in the midst of waters."

In the early days of the church in Europe, spots for monasteries were often picked for how close they were to good fishing spots, and the monks built fishponds as a backup for the days when they fasted from meat. The first essay in English on sportfishing, "Treatise of Fishing with an Angle," was published in 1486 by Dame Juliana Berners, a Christian nun who fished with a three-piece, twelve-foot-long rod and fifteen-foot tapered fly line, with flies dyed in different colors, and tied of wool and partridge and duck feathers.

There is an amazing theory about the ancient origin of the connection between fish and Jesus. You could call it "The Jesus Code." The Greek word for fish is

Ichthys, and according to the theory, this is code for the Greek words for "Jesus Christ, Son of God, Savior," which is "Iesous Christos Theou Yios Soter," or I-C-Th-Y-S. Pretty convincing, isn't it?

According to Luke:24, Jesus Christ's last meal on Earth was fish. "Peace be unto you," he said to his amazed disciples, and asked, "Have ye here any meat?" They gave him "a piece of a broiled fish" and a piece of honeycomb. After he enjoyed the fish, Jesus blessed them, and was then "carried up into heaven."

The spiritual and symbolic power of fish goes beyond Christian cultures, to belief traditions that have also made their way to America, too.

In the Jewish tradition of a Rosh Hashanah feast, a fish or goat head is positioned on the table, because Rosh Hashanah means "head" of the year. In Asian traditions, fish can symbolize marriage, love, courage, abundance, and prosperity. In Buddhism, the fact that fish don't close their eyes reflects the concentration and awareness of a spiritually dedicated person. Wooden fish are used in Buddhist ceremonies to lead the rhythm of chanting. In the myths of Hawaii, every island has its own shark god. In the tradition of the Kwagiulth people of Western Canada, the first human to appear after the great flood emerged from under the skin of a halibut.

Let the waters swarm with fish and other life.

—GENESIS 1:20A (NEW LIVING TRANSLATION)

No life is so happy and so pleasant as the life of the well-govern'd angler.

—IZAAK WALTON

Time is but the stream I go a-fishing in. I drink at it; but while I drink I see the sandy bottom and detect how shallow it is. Its thin current slides away, but eternity remains.

—HENRY DAVID THOREAU

The best part of hunting and fishing was the thinking about going and the talking about it after you got back.

—ROBERT RUARK, *THE OLD MAN AND THE BOY*

The fisherman knows many things.

He knows the sublime patience of solitude in the wilderness, of fly-fishing for supper in a cold, clear stream.

He knows how to "read the water" and "fish the seams," how to see the river as a living creature that changes with every moment and carries a hungry trout in endlessly changing currents, lanes, and pockets between the fast water and slow water.

He knows that many fish don't swim upstream, they swim into the current, and that this determines how you should present your fly.

He knows lots about bugs, and how to poke around his fly box to "match the hatch" and pick just the right pattern to snag a crafty bass.

She knows how to spend summer afternoons in her backyard practicing casting for hours and hours, so her moves on the water will be graceful and precise.

She knows how, before casting, to carefully plan out in her mind how she'll hook, fight, and land the fish, and she knows how slowly to move her feet while wading out in a still pond in search of cagey trout.

She knows how to draw a secure knot, moisten it with water or saliva, and adjust the number of turns to keep it from slipping.

He knows, in the words of legendary fisherman Lefty Kreh, that "The angler who fishes carefully along the stream bank in deep shade is much less likely to be seen by the fish than the angler working in sunlight. By approaching from the shaded side you'll reduce the contrast between yourself and the bank, lessening the chances of being seen by the fish."

She knows how to read the water so well on shallow saltwater flats that she can identify what type of fish is making a tiny, telltale shaking movement on the water surface.

She knows how to gently ease the lure along a river, lock her thoughts into the handle and down the rod and line, and "finger-feel" as the lure brushes along rocks, gravel, weeds, and sand, wherever the bass is lurking.

He knows that when obstacles or currents make a downstream presentation impossible, how to make a "curve cast" with a powerful forward cast of the fly line, then stopping it quickly so the leader connection hinges and shoots the fly sideways to the target. He knows how to "flip" a gentle underhand cast to place a lure in thick cover or dirty water.

He knows that 20 percent of the fishermen catch 80 percent of the fish, and as professional fisherman Roland Martin put it, "Every fisherman out there on the lake is going to have some lucky opportunity come by, but half of them aren't going to recognize the luck. And half of them aren't going to capitalize on the luck. Really good fishermen capitalize on those opportunities."

And finally perhaps, as he fishes, he knows somewhere in the back of his mind that he is spiritually connected to millions of other American Fishermen who have shared his passion over the course of the past five thousand years.

From the days of the Native Americans, through the rise of the modern United States, fishermen helped found, feed, and forge America. Fish have always been close to the heart of the nation, as sustenance, as business, as sport, and as spiritual inspiration. Fish helped feed and save the first European settlers in the New World. Fish provided a critical source of income for the Founding Fathers of the United States, they powered the epic westward expansion of the nation, and they affected American history at certain surprising and long-forgotten turning points. The act of fishing helped refresh and inspire a number of our presidents as they grappled with the destiny of the nation. And for millions of Americans, the sport of fishing has provided a powerful source of health, strength, relaxation, and renewal.

It all began, as many American stories do, with the First Americans, the native peoples.

1
The Founding Fishermen

Only when the last tree has been felled, the last river poisoned, and the last fish caught, man will know, that he cannot eat money.

—NATIVE AMERICAN SAYING

Of all places on earth, America is among the lands that are most blessed with creatures of the water, and for thousands of years, fishing has been part of the American soul.

From the coast of Maine to the Gulf of Mexico, from the rushing rivers of the Rocky Mountains and the Catskills to the waters of Chesapeake Bay, Alaska, and Hawaii, America has always enjoyed a stunning legacy and diversity of fish. By some estimates, the preco-

Native American spearfishing in northern California.
(Edward S. Curtis/Library of Congress)

lonial sizes of the salmon populations in the Pacific Northwest and Atlantic regions of the continent were well over 10 million each, and as recently as the early twentieth century, the population of cod in the Grand Banks and Georges Bank zones of the North Atlantic were estimated as over one billion.

For the first Americans, the native peoples, the waters in and around the continent were the source of an endless supply of fish, and for many centuries before the arrival of Europeans, fish were a critical source of nourishment and survival for a number of native cultures.

To the Native Americans, the continent was a fantastic cornucopia of every kind of fish imaginable, and in some spots you could practically cross a lake or river by stepping on fish. Offshore tuna flocked around both coasts, the banks off New England were home to gigantic schools of cod and haddock, and Atlantic salmon flourished all along the East Coast. In the Arctic, Thule Inuits in skin-lined boats stalked bowhead and beluga whales with bone-tipped harpoons.

Along the eastern and Gulf coasts were huge populations of striped bass, and in spawning season, the eastern rivers and streams down to Florida were choked with shad. Pacific salmon ruled the coasts, estuaries, and rivers from Alaska down toward California, flanked by clams, smelt, herring, and anchovies. Brook trout (not really a trout but a char) frolicked in wilderness rivers and lakes as far south as Georgia and as far west as Iowa and Minnesota. Farther west were the cutthroat and rainbow trout and bull trout (also a char), and the ocean-trekking steelhead.

The Native Americans developed a host of ingenious ways to capture fish—including hand lines, buffalo-bone hooks, spearfishing by torchlight, bow-and-arrows connected to lines, throwing-sticks, lures, decoys, poles, canoes, natural barricades, seine nets, intricate trapping systems, and even their bare hands for grabbing slow-moving shallow-water catfish.

One "fish weir," an elaborate trap complex used by Native Americans near today's downtown Boston as long as five thousand years ago, consisted of 65,000 wooden stakes pounded into a three-acre area of shallow mudflats on the tidal shoreline, interlaced with a web of tree branches that allowed water to pass through, but captured the fish. A similar ancient trap discovered in Sebasticook, Maine, featured a small opening that enabled eels and fish to swim upstream to spawn, and trapped them on the downstream voyage.

In Virginia, Native American tribes fashioned V-shaped traps from stones along the James and Potomac Rivers. An early colonial historian named

Native Americans fishing in the rapids, Sault Ste. Marie, Michigan, circa 1900. (Library of Congress)

Robert Beverley Jr. described the technique in 1705: "At the falls of the rivers, where the water is shallow, and the current strong, the Indians use another kind of weir. . . . They make a dam of loose stone where of there is plenty on hand, quite across the river, leaving one, two or more spaces or tunnels, for the water to pass through at the mouth of which they set a pot of reeds, wove in form of a cone, whose base is about three foot, and in perpendicular ten, into which the swiftness of the current carries the fish, and wedges them in fast, that they cannot possibly return." In some native tribes, a whole clan or village would pitch in to

Native American fisherman using a dip net, Washington state, 1910. (Edward S. Curtis/Library of Congress)

Native American fishermen in a dugout canoe, Virginia, sixteenth century. (Theodor de Bry/Library of Congress)

scoop up the fish in nets and baskets, then dry and salt or smoke the fish for storage, or bury it to enrich the soil for planting crops.

A thousand years ago in the American Southwest, in what is now New Mexico, the Mimbrenos Native Americans not only fished in the Gila and Mimbres Rivers with traps, nets, and fishing lines—they appear to have gone on fishing expeditions as far off as the Gulf of California (also known as the Sea of Cortez), hundreds of miles away.

How do we know this? They left behind a "prehistoric photo album" of more than 10,000 pieces of pottery, many illustrated with artwork that features fishing scenes and portraits of pike, gar, trout, catfish, snapper, angelfish, and swordfish. One piece shows a fisherman wearing a giant bird's head and holding up a big fish in one hand and a string of eight smaller fish in the other. He looks just like a modern fisherman proudly showing off his catch. Who knows, it may even have been a "selfie"!

In Arctic regions, Inuit fishermen hunted whales and seals with hand-thrown, deep-penetration harpoons connected to pull-lines. In the Gulf of Mexico and off the Pacific Coast, a dugout canoe could hold twenty or more fishermen. In the Pacific Northwest,

Native Americans built riverside platforms from which to spear jumping salmon.

In the waters surrounding the Hawaiian Islands, native peoples hunted ocean and river fish with nets, spears, bonehooks, lures, canoes, live bait, lines made of plant fibers, and stone sinkers called *pohakialoa,* which dropped lines to far depths. Whole villages would get together for a team fishing event called a *hukilau,* when everyone manipulated a giant net to scoop up shallow-water fish just offshore, which they split up equally.

Native Hawaiian fishermen developed an amazing process for capturing shark. For several days, they tossed bait in shallow water from their canoes, along with narcotic black *awa* root. After a few days, the happy shark, increasingly gorged on food and stoned on *awa,* came close to the boats for more treats, and instead found they had a noose around their groggy heads, and were being towed to shore. Some fishermen became incredibly talented in handling shark. The nineteenth-century Hawaiian historian Samuel Kamakau wrote, "To the native son, the shark was a horse to be bridled, its fin serving as the pommel of a Mexican saddle. I have seen men skilled in herding sharks riding a shark like a horse, turning the shark

Hawaiian fisherman with net.
(Charles Bartlett, 1920/Library of Congress)

to this side and that until carried to shore, where the shark died."

When an English naturalist named William Bartram tagged along on a Native American fishing trip along present-day St. Johns River in Florida in 1773, he watched expert anglers use long rods, fishing line, hooks, and lures to snag trout along the edges of floating insect clusters. They were, in other words, fly-fishing.

Bertram watched spellbound as the master anglers expertly plied their trade. "They are taken with a hook and line, but without any bait," he recalled. "Two people

Leif Erickson spots America.
(New York Public Library Digital Collections)

are in a little canoe, one sitting in the stern to steer, and the other near the bow, having a rod ten or twelve feet in length, to one end of which is tied a strong line, about twenty inches in length, to which are fastened three large hooks, back to back. These are fixed very securely, and covered with the white hair of a deer's tail, shreds of red garter, and some parti-coloured feathers, all of which form a tuft or tassel, nearly as large as one's fist, and entirely cover and conceal the hooks: this is called a bob. The steersman paddles softly, and proceeds slowly along shore, keeping the boat parallel to it, at a distance just sufficient to admit the fisherman to reach the edge of the floating weeds along shore; he now ingeniously swings the bob backwards and forwards, just above the surface, and sometimes tips the water with it; when the unfortunate cheated trout instantly springs from under the weeds, and seizes the supposed prey. Thus he is caught without a possibility of escape." The native fisherman angler quickly yanked the fish into the canoe.

For thousands of years, America was a secret Paradise of Fishing, known only to the native peoples.

Around the year A.D. 1000, all that was about to change.

Fish brought the first Europeans to America.

In fact, modern America was born, in large part, because of a tough, tasty, peculiar fish called *Gadus morhua,* otherwise known as the Atlantic cod.

The first Europeans to leapfrog across the North Atlantic to Iceland, Greenland, and North America were the Norwegian people known as Norsemen, or, as their name became when they were raiding and pillaging, the Vikings.

In about the year 980, when historians now believe Norse Greenland settler Leif Erikson began the first try at a European-American settlement at present-day Newfoundland, the 160 or so colonists who followed in at least five expeditions to America between 985 and 1011 managed to survive, just barely, in part because of Atlantic cod.

For a globe-trotting band of marauders like the Vikings, cod was the perfect trail-and-travel superfood, packed with vitamins and protein. The fish was easy to catch in shallow waters of 120 feet (20 fathoms) or less. It exerted its muscles so little that its color was white, and its blood contained antifreeze, which enabled it to flourish in supercold sub-arctic waters. It swam with its mouth open, gobbling up whatever came in, including baby cod. You can eat nearly all parts of the cod: skin, cheeks, head, throat, stomach, tripe, liver, tongue, air bladder, even the sperm and female private parts if you're so inclined.

Best of all, the fish bred in staggeringly large numbers all across the offshore continental shelf areas and banks that linked the Vikings' travels across the North Atlantic in sail-and-oar-powered open longboats, enabling them to hop between continents, something no explorers are known to have done before. Until then, long sea voyages beyond coastal regions were impossible since provisions quickly spoiled at sea, but according to Mark Kurlansky, author of the bestseller *Cod: A Biography of the Fish That Changed the World*, the Vikings had learned to "preserve codfish by hanging it in the frosty winter air until it lost four-fifths of its weight and became a durable woodlike plank." You could break off a piece and warm it up in hot water or just chew on it like hardtack.

The Norsemen loved the tiny stretches of North America they managed to explore, which they called Vinland. "There they found self-sown fields of wheat where the ground was low-lying and vines wherever it was hilly," said a Norse saga. "Every brook there was full of fish and when the tide went out, there were halibut in the trenches. There were vast numbers of animals of every kind in the forest."

At first, there was nothing to raid or pillage, so the Norsemen settled down as farmers and fishermen at a spot on the Newfoundland coast at present-day L'Anse aux Meadows. They met local native people and traded red cloth, iron, and milk for furs. But the experimental colony was doomed. Disputes erupted, and the Europeans were vastly outnumbered with no hope of reinforcement, their axes and spears no match for the native's bows and arrows.

During one skirmish, Leif Erikson's brother Thorvald Erikson was hit straight through the belly by an arrow, which he pulled out by hand. As he died, he allegedly uttered a fitting epitaph for the Vikings

The humble codfish, upon which the foundations of early America were built. (New York Public Library Digital Collections)

in North America: "We have won a fine and fruitful country, but will hardly be allowed to enjoy it."

After a few years of brutally harsh winters and skirmishes with the locals, the Norse settlers abandoned the micro-colony, which lay undiscovered beneath the Newfoundland soil until archeologists dug up its remains in 1960.

But beyond the foggy eastern horizon, a secret was unfolding, a secret the Vikings had no way of knowing.

At possibly the exact same time the Norse colony was failing, a new breed of American fishermen was quietly scooping up vast amounts of Atlantic cod across the area now called the Grand Banks, off present-day Newfoundland.

They were from a tiny nation of highly skilled anglers from northern Spain. They were called the Basques.

The Basques were master whalers and fishermen who are believed to have discovered the northern Atlantic treasure trove area of cod perhaps as early as the year A.D. 1000. They one-upped the Vikings by adding a secret ingredient to the cod—salt.

Salt curing was better than Viking-style air curing, because it enabled the cod to be shipped in bulk, without spoiling, for much longer times and distances. This meant the Basques could sell it in the newly fish-crazed markets of Europe, where meat eating on many days of the year was banned by the medieval church. When the flaky product was soaked, cooked, and flavored in a European kitchen, it tasted even better than fresh cod.

A highly profitable fishing business empire boomed for many years, and it is speculated that the Basque fishermen probably set up fish curing stations along North American coastlines, though no trace of them has ever been found. But like any good fisherman zealously guarding the secret of a choice fishing spot, the tight-lipped Basques kept the treasure of Atlantic cod secret. In Europe, rumors swirled that the Basques had discovered a land beyond the western horizon, but they never told anyone exactly where all the fish came from.

Finally, on June 24, 1497, the secret was blown, when a Genoese explorer named Giovanni Caboto, otherwise known as John Cabot, discovered the cod-rich Grand Banks while on a mission for Henry VII of England. A diplomatic letter later that year marveled, "The sea there is swarming with fish which can be taken not only with the net but in baskets let down with a stone."

A "Cod Rush" kicked off, which triggered English colonization of North America. The Basques held on to their American fishing grounds as long as they could, and as late as 1534, a witness reported spotting one thousand Basques around the St. Lawrence River and Hudson Bay, still fishing away.

To look into the depths of the sea is to behold the imagination of the Unknown.

—Victor Hugo

The pleasant'st angling is to see the fish
Cut with her golden oars the silver stream,
And greedily devour the treacherous bait. . . .

—William Shakespeare, *Much Ado About Nothing*

The water you touch in a river is the last of that which has passed, and the first of that which is coming; thus it is with time.

—LEONARDO DA VINCI

When the Pilgrims first came to America in the early 1600s, they thought about fish. A lot.

This gave them a big idea.

"We'll starting a cod-fishing business," they figured. "We'll do some hunting and fishing. We'll eat well, worship our God freely, and we'll make a fortune!"

The plan made perfect sense. New England had a year-round fishing season. It was near the richest cod banks in the world. The New World was, you could say, their oyster.

What could go wrong? Plenty, as it turned out.

The Pilgrims had forgotten to pack much fishing tackle on their first trips across the Atlantic. They had no idea how to fish, and apparently none of them had studied up on the subject, either. They knew even less about hunting and farming. Within months, fully half of the one hundred or so Pilgrims who landed on what is today Provincetown, Massachusetts, at the tip of Cape Cod on November 11, 1620, were dead from sickness and starvation.

The Pilgrims land at Plymouth Rock in 1620. They'd hoped to start a cod-fishing business, but they forgot to bring much fishing tackle. Plus, they had no clue how to fish. Pretty soon, they were starving. (Currier & Ives/New York Public Library Digital Collections)

When it came to seafood, the Pilgrims were, at first, squeamish. They were reportedly disgusted by eels and mussels. They watched Native Americans spear nearly six-foot-long sturgeon in the streams, harvest clams on the shoreline, and use bonehooks and nets and line fashioned from vegetable fibers to catch various fish, but at first had little interest in fishing as a main source of food. The Pilgrims were instead focused on trying to catch offshore cod for sale, but they had little luck.

In 1620, the Pilgrims just barely managed to pull in their first harvest, so they threw a party, or "Thanksgiving" celebration. The ninety or so invited guests of the native Wampanoag tribe, whose home the area had been for the previous ten thousand years or so, are believed to have included seafood in their party gifts, possibly in the form of items like locally harvested eel, clams, cod, sea bass, and shellfish.

By the spring of 1622, when the Pilgrims were again starving to death, they sent a ship over to the small English fishing station on Damariscove Island, in what is now Maine, to beg for food. The English fishermen there filled their baskets with cod, enough to enable the Pilgrims to survive. Not only were the Pilgrims good fish beggars; they were good scavengers, too. During their first terrible days of near starvation in America, they also survived by swiping stashes of food that had been buried by natives. Eventually the Pilgrims got smart, welcomed the help and advice of native fishermen, and learned how to catch fish, lobster, and eel, and how to bury local fish in the soil to fertilize crops.

The Pilgrims learned that the waters in and around New England were rich with all kinds of fish, including the succulent bass. As one enthusiast later wrote of the striped bass, "The bass is one of the best fishes in the country and though men are soon wearied with other fish, yet are they never with bass; it is a delicate, fine, fat, fast fish, having a bone in his head, which contains a saucerful of marrow sweet and good, pleasant to the pallate, and wholesome to the stomach. When there be great store of them we only eat the heads and salt up the bodies for winter." He added, "Of these fishes some be three and some four foot long, some bigger, some lesser. At some tides a man may catch a dozen or twenty of these in three hours."

Early New Englanders had so much fish to choose from that they got very picky. Halibut, for example, was looked down on. One colonist reported, "The halibut is not much unlike a plaice or turbot, some being two yards long and one foot wide and a foot thick; the plenty of better fish makes them of little esteem except the head and fins, which stewed or baked is very good. These halibuts be little set by while bass is in season. Thornback and skate is given to the dogs, being not counted worth the dressing in many places."

Lobsters later became an American delicacy, and the first Pilgrim settlers ate them as survival food, but in New England in the 1600s there were so many of them that they too were scorned by the European settlers and left to the natives. One colonist reported, "The Indians get many of them every day to bait their hooks and to eat when they can get no bass." The native fishermen caught fish and shellfish close to shore, in eight-man boats made of birch bark that could only travel short distances. But there was so much to catch near the shore, according to marine scientist Callum Roberts, that they had "no need to fish in deeper water farther from land."

By 1623, the Pilgrims tried to set up a commercial fishing station in Gloucester, but it tanked. Two years later they tried again, and failed. They sent word to England that they needed some serious fishing tips and gear, and eventually they became commercial fishermen. Mighty good ones, too. They set up a network of fishing posts around New England, which thrived. At Provincetown and Gloucester, windmills churned ocean water into evaporation pools to create salt for fish curing.

The settlers soon mastered the process of harvesting river fish, too, using weirs, or wooden traps, inspired by those of the Native Americans. A French visitor to the Plymouth Colony in about 1628 described one impressive system: "At the south side of the town there flows down a small river [probably Billington Brook] of fresh water, very rapid, but shallow, which takes its rise from several lakes in the land above. . . . This river the English have shut in with planks and in the middle with a little door, which slides up and down, and at the sides with trellice work through which the water has its courses but which they can also close with slides. At the mouth they have constructed it with planks, like an eel-pot, with wings, where in the middle is also a sliding door and with trellice work at the sides, so that between the two dams there is a square pool into which the fish aforesaid come swimming . . . in order to get up above."

By 1640, the Massachusetts Bay Colony, headquartered at Boston and run by the more powerful Puritans, was shipping 300,000 cod to the global fish markets. "Under the laws of the British Empire, colonists were supposed to have sold their cod to England, but they were trading directly with Spain and the Caribbean," wrote Mark Kurlansky. "When the British Crown complained in 1677, Charles received 1,000 codfish and a polite note from the New Englanders saying they would do what they liked. It was the beginning of their independence." Those New Englanders always were a feisty bunch! Later, he added, "when the British crown became concerned about New England having gotten out of control and they wanted to control the codfish trade, they did it by the Molasses Act, by putting a tariff on molasses, they were indirectly restricting cod trade, because it was cod for molasses."

By the 1700s, New England was a global fishing superpower, and several families amassed gigantic fortunes as members of the "Codfish Aristocracy." The descendants of the outcast, starving Pilgrim Fathers, who had sailed on the *Mayflower,* were building grand estates and giant mansions decorated with cod designs. The commercial whaling business boomed, too, triggering industrial development in New England and beyond.

There was, however, a terribly dark side to the fish-powered transformation of New England. It was based on a deal with the Devil himself—in the form of the flourishing global traffic in human slaves.

Many of us were taught about the "triangle trade," the three-way Atlantic traffic in slaves, molasses, and rum that powered early America. But what's much lesser known is the fact that this process included a fourth commodity—cod.

It began in 1645, when a New England ship journeyed to the Cape Verde Islands, bought enslaved Africans, took them to market in Barbados, and went back to Boston with various Caribbean products. Pretty soon, New England salt cod was added to the mix. "By the mid-17th century, ships from Boston were delivering salt cod to Spain and Portugal," wrote author Malabar Hornblower. "From there they moved on to West Africa, where they bought slaves. These they transported to the West Indies and bartered for sugar and

Collect Pond in New York City, the birthplace of sportfishing in colonial America. (Kenneth Holcomb Dunshee/City University of New York)

molasses. Finally, coffers bulging, they made their way back to the harbors of New England—and the rum distilleries." The lowest-quality cuts of salt cod were called "West India cure," and it was a cheap, high-protein food used to feed slaves working as sugarcane cutters and boilers in the plantations of the Caribbean.

The profits on this fish-based trade in slavery were huge, and helped create the commercial foundation of colonial America.

The next time you're in New York City, take the number 6 subway to Chinatown's Canal Street, then walk three blocks south on Lafayette Street. When you get to the corner of Franklin Street, look to your left.

You'll see a tiny half-forgotten park with a little reflecting pool, flanked by the steel and concrete jungle of Manhattan. It's called Collect Pond Park. Most folks don't know it, but this marks the exact spot where a great American sport was born in colonial America—recreational sportfishing.

The great American pastime of sportfishing was born here as early as 1630, when people in the then-Dutch settlement of New Amsterdam organized recreational fishing parties at what was then a bucolic, much larger 48-acre, 60-foot-deep fishing hole called *Kolch,* or "small body of water."

Back then, the island of Manhattan was largely a solid forest, with streams and lakes that were home to multitudes of frolicking fish, including brook trout in Collect Pond. It was a beautiful place to fish, nestled in a valley beside Bayard Mount, which at 110 feet was the highest hill in southern Manhattan. After the English taking of New Amsterdam in 1664, the name was changed to "Collect Pond" and was later known as "Fresh Water Pond."

For many decades, Collect Pond was New York's favorite fishing spot. It got so popular that in 1734 the authorities had to protect the brook trout population

Early American fishing scene. (Library of Congress)

by banning commercial fishermen from the pond and limiting it to recreational fishing only. Eventually the pond got too filled with garbage to survive, but what began at Collect Pond rippled out across the land, and recreational fishing eventually became America's biggest sport.

Some of the biggest sportfishing enthusiasts in early America were British army officers stationed in the colonies. They enjoyed the already sophisticated sport of English fly fishing in the rough, pristine waters of Pennsylvania's hinterland, as well as the rivers and lakes of New York's Long Island. By the mid-1700s, recreational fishing was flourishing in the rich trout waters of Southern Pennsylvania. The first American fishing club, the Schuylkill Fishing Company, popped up in 1732 on the banks of the Schuylkill River in Philadelphia, soon joined by four others in the area.

By 1776, Philadelphia already had several sporting tackle shops, selling rods made of dogwood, bamboo, and hazel, and fine Kirby hooks imported from London. The first superstar fly-tier in the colonies at the time was a Pennsylvania Quaker named Hugh Davis, who was renowned for his exquisite feathered flies and sturdy deep-sea tackle.

Of course, that year also saw the birth of the United States of America itself.

It was an event made possible by one of the greatest American Fishermen of all time.

2
The Fisherman Who Created America

I again repeat that when the Schools of fish run, you must draw day and night.

—President George Washington

George Washington was many things—brilliant general, two-term president, duckhunter, fox-hunter, world-class horseman, one of the tallest men in colonial America, farmer, slaveholder, visionary entrepreneur, and father of the United States.

He was also a master American Fisherman.

In fact, if it wasn't for fish, there might never have been a President George Washington. Or even, for that matter, a United States.

Washington's interest in fishing dated back to his youth, when he learned the ways of country life, including fishing, on his family estate at Mount Vernon, Virginia, and later when he lived off the land as a young militiaman on the American frontier.

When he was nineteen years old, George Washington pioneered an American sport that became hugely popular two hundred years later. He took a cruise to the Caribbean and did some deep-sea fishing. That's right. The father of our nation was also one of our first saltwater ocean fishermen!

It was the fall of 1751, and Washington was traveling with his sickly older half brother, Lawrence, on a trip to Barbados to soak up the sunshine and fresh ocean air, which they hoped would help restore Lawrence's health. They hitched a ride on a small trading sloop called the *Success,* on a six-week trip that sometimes got so rough, Washington wrote in his travel diary, that the Atlantic Ocean was "fickle and merciless."

When the waves calmed down, the two brothers realized that the sea around them was teeming with dolphin, pilot fish, shark, and barracuda, so they decided to throw a hook and line overboard and try their luck at some good old-fashioned saltwater angling. On October 7 they had a long, action-packed day of fishing that George described in his diary: "a dolphin we

America's Founding Fisherman. (New York Public Library Digital Collections)

catched at noon but could not entice with a baited hook two Baricootas, which played under our stern for some hours; the dolphin being small we had it dressed for supper."

Once the brothers landed in Barbados, George became enraptured by the island, as many visitors do. His diary entry of November 6, 1751, revealed how much he enjoyed the first trip away from the American continent: “In the cool of the evening we rode out accompanied by Mr. Carter to seek lodgings in the country, as the Doctor advised, and were perfectly enraptured with the beautiful prospects, which every side presented to our view, the fields of cane, corn, fruit-trees, in a delightful green.”

Unfortunately, the excellent adventure soon turned sour, when George contracted smallpox and Lawrence’s tuberculosis didn’t improve much. But the experience sharpened George Washington’s fishing skills, which eventually gave him the cash he needed to finance the spectacular political and military career that followed.

For sixteen years before the American Revolution, George Washington ran his inherited family estate of Mount Vernon plantation as the CEO of a bustling, diversified, global agro-industrial enterprise based on two pillars: farming and fishing. Washington’s commercial fishing operation was the most consistently profitable segment of his portfolio, bailing him out of a number of years when his other businesses bled red ink. Various business deals and crops came and went, but every year, the fish always came through for George Washington.

At Mount Vernon, Washington raised cattle, processed the meat, and marked it with the “GW” brand.

He made boots and shoes, ran a cargo-carrying schooner, manufactured textiles, and mass-produced whiskey. He speculated in real estate and expanded his homestead from 2,000 acres to 8,000. He grew tobacco, which chewed up too much farmland, so he switched to wheat and corn. Washington's mill ground 278,000 pounds of branded flour a year, and he shipped merchandise to buyers throughout the American colonies, England, and Portugal. He was a micromanager who kept a close eye on all the details of his businesses, haggled with buyers and suppliers over prices, and threatened lawsuits over deals gone bad.

But it was as a commercial fisherman that Washington really made his mark as a businessman. In fact, he was the only president who made a major part of his living from commercial fishing.

While Washington's crops sometimes failed, there was one source of cash that always delivered the goods—fish. Mount Vernon was nestled on the banks of the mile-wide Potomac River, which, Washington wrote in 1793, was "a river well-stocked with various kinds of fish in all seasons of the year, and in the spring with shad, herring, bass, carp, perch, sturgeon, etc. in great abundance."

Washington set up a full-scale fish catching and processing facility on his estate, an operation that handled

George and Martha Washington with the French general Marquis de Lafayette at the entrance to Mount Vernon. Their opulent lifestyle was financed, in large part, by fish caught in the Potomac River, on the right. Inside the mansion, broiled fish was often on the menu. (Library of Congress)

everything from hauling in the catch to packing it in barrels for long-distance shipping. He and his wife, Martha, regularly served broiled fish to their guests at Mount Vernon.

There is no sugarcoating an ugly truth: Washington's lucrative fishing operation was the product of human slave labor. During the spring spawning

Mount Vernon, headquarters of George Washington's artisanal fish-processing and agribusiness. Every spring, Washington's slaves netted thousands of fish in the Potomac River (behind the mansion). His fish brand acquired a global reputation for excellence and quality, and financed his rise to power. (MountVernon.org)

period of March, April, and May, scores of Washington's black slaves were pressed into service to perform the backbreaking work of rowing a small fleet of boats far out into the river, throwing out giant nets that stretched up to five hundred feet long, hauling in the thousands of fish, building the shipping barrels, and cleaning, gutting, salting, and packing the fish for sale.

The work became frantic at the peak of the season. "I again repeat," Washington commanded the slave overseer, "that when the Schools of fish run, you must draw day and night." The slaves worked in shifts late into the evening, and some of the biggest catches were done by torchlight.

If they did a good job hauling in the catch, Washington rewarded his slaves with an extra ration of liquor. One visitor noticed a strict racial separation of the catch that was eaten at Mount Vernon; the white catfish and perch was fit to be eaten by the whites, but the gar and another type of catfish, "which is black, is left for the blacks." Washington also fed his slaves salted herring, and he wrote that fish made up "the larger part of the flesh diet of my people." The leftovers and inedible parts were used for fertilizer.

Washington's artisanal fish products were so good that the Mount Vernon brand won a global reputation for excellence. He packed fish for sale at the markets in nearby Alexandria and exported them to Antigua and the West Indies, where they were often used to feed slaves.

For the father of our nation, fish was big business. In 1771, for example, he caught 7,760 American shad and 679,000 herring, and they were money in the bank for George Washington. A visitor to Mount Vernon in 1798 reported that over 100,000 herring were hauled ashore in a single day.

Washington's fish enabled him to build his growing reputation as a Virginia gentleman and finance his budding career as a politician. If it weren't for fish, Washington probably would have gone bust many years before he led the Continental Army to victory, or at least chosen another career. Ironically, fish helped trigger both the American Revolution, and indirectly the Civil War, too, since the Revolution was launched as a reaction to British interference in the molasses-salt-fish trade between New England and the West Indies—and the demand by Massachusetts for fishing rights to the Grand Banks fueled some of the first tensions between the North and the South.

In the darkest hours of the American Revolution, at the very moment when the fate of the infant nation hung in the balance, a group of saviors arrived in the distance.

In one of the most incredible "creation stories" of American history, the rescue force was made up of thousands and thousands of tasty, nutritious fish.

Deep in the savage winter of 1777–78, Washington's troops were on the brink of freezing and starving to death. Most of the Continental Army was camped out in heavy snow, improvised huts, and the open woods and fields at Valley Forge, Pennsylvania—12,000 men—some reduced to wearing scraps and rags, and all freezing and constantly running out of food. Their desperate commander, General George Washington, begged America's governing body during the Revolution, the Continental Congress, for food. He asked the friendly local governors for food. He pleaded with every local merchant, butcher, and farmer he could find. But he kept running out of food.

In theory, by an act of the Continental Congress, every American soldier was supposed to get a minimum daily ration of a pound of meat or fish, a pound and a half of bread, a serving of whiskey or spirits and, when available, a half pint of peas or beans. But in practice, they often went hungry.

On December 22, 1777, a helpless, furious Washington wrote to the Congress, "I do not know from what causes this alarming deficiency or rather total failure of supplies arises, but unless more vigorous exertions take place immediately, the army must dissolve."

The American troops were fast running out of hope, and Washington feared they were on the verge of mutiny. On February 16, Washington wrote to Governor George Clinton of New York, "For some days past, there has been little less than a famine in camp. A part of the army has been a week, without any kind of flesh [to eat], and the rest for three or four days. Naked and starving as they are, we cannot enough admire the incomparable patience and fidelity of the soldiery, that they have not been this excited by their sufferings, to a general mutiny or dispersion. Strong symptoms, however, [of] discontent have appeared in particular instances; and nothing but the most active efforts everywhere can long avert so shocking a catastrophe. Our present sufferings are not all. There is no foundation laid for any adequate relief hereafter. All the magazines provided in the States of New Jersey, Pennsylvania, Delaware and Maryland, and all the immediate additional supplies they seem capable of affording, will not be sufficient to support the army more than a month longer, if so long."

In February 1778, soldiers were chanting, "No pay! No clothes! No provisions! No rum!" By March, Washington recalled, "rations dwindled and soldiers were driven to scavenge," and "morale ebbed," "with little hope of finding food." He later wrote that "no history now extant can furnish an instance of an army's suffering such uncommon hardships as ours has done," "without clothes, blankets, shoes, or provisions," "marching through the frost and snow."

But a vision of salvation appeared in the distance. Valley Forge, as it happened, was located on the banks of Pennsylvania's Schuylkill River, and the river was home to throngs of shad who migrated every year from Delaware Bay through Philadelphia and up the

Schuylkill to Valley Forge, right past Washington's camp. As a commercial shad fisherman, Washington would have known this. But as of March 5, the fish weren't due to arrive for at least a week or two. Would they make it in time to save the United States of America?

Then, it seemed, a miracle occurred.

As historian Harry Emerson Wildes told the story in his 1938 book, *Valley Forge,* "Then, dramatically, the famine completely ended. Countless thousands of fat shad, swimming up the Schuylkill to spawn, filled the river. Soldiers thronged the river bank. The cavalry was ordered into the river bed. The horsemen rode upstream, noisily shouting and beating the water, driving the shad before them into nets spread across the Schuylkill. So thick were the shad that, when the fish were cornered in the nets, a pole could not be thrust into the water without striking fish. The netting was continued day after day until the army was thoroughly stuffed with fish and in addition hundreds of barrels of shad were salted down for future use."

Fish to the Rescue! The Continental Army troops chowed down on shad, regained their strength, and went on to beat the toughest army in the world! America was saved, just in the nick of time! Ever since Wildes wrote his book, the story made its way into countless other books and articles about Valley Forge.

It's a wonderful story—but there's one slight problem. You see, it may not have happened, at least not the way Mr. Wildes told it. A few years back, historians did a good deal of research on the subject and discovered there were no historical records to support the claim that the shad came through in such a dramatic fashion. In fact, there is evidence the British stretched a seine across the Schuylkill River near Philadelphia, specifically to *block* the shad from Washington's troops. Other details of the story didn't check out, either.

But the story does contain the morsel of an important truth—barreled salt shad, pickled herring, and other fish were indeed a regular provision for Washington's troops during the Revolution, and offered a critical source of nutrition for the soldiers at several points during the epic war for liberation, including the ordeal at Valley Forge. Salted fish was a nearly perfect food choice for Washington and his men, though they usually preferred meat when they could get it. It was nourishing and convenient, as it could be shipped and stored long distances without refrigeration.

The story of how the American Revolution was saved at Valley Forge was indeed partly a story of fish. In

January and February 1778, for example, some 35,000 pounds of herring was issued at Valley Forge, and in one seven-day period in May of that year, around 22,000 pounds of fish made it to the chow line.

On February 19, when it looked like the whole army was about to starve, an emergency alert was issued to patriots in Virginia and Maryland to round up 18,000 barrels of fish, preferably shad, and express-deliver them to Valley Forge. The next month, another 1,000 barrels of shad were ordered. Weeks later, in early April, letters went out from Valley Forge to "procure a large quantity recommend Shad Fish only" and to "put up all the Fish you possibly can."

On April 8, a message went out from a soldier at Valley Forge to an associate in New Jersey that instead of serving with the militia, "you can render ten fold more service to your country by paying proper attention to the fisherys in your neighbourhood. All the tight barrels in camp shall be sent to you, let no opportunity be lost to procure all the fish you can and be very particular in salting them."

Washington's supply officers sent out messages on April 29 and 30: "What success have you had in procuring shad? Hope pretty considerable," and "Send to headquarters two brigades of teams [roughly twenty-four supply wagons] loaded with herrings, shad

Washington at Valley Forge during the savage winter of 1777–78. His men are on the brink of starvation, and he is afraid they are about to mutiny. (New York Public Library Digital Collections)

and butter. Pray exert yourself in the fisheries and providing Teams to forward all the Stores in your Neighbourhood." From May 25 to 31, some 15,508 pounds of fish were issued to the troops.

As recruits streamed into camp in April and May, and additional regiments arrived from Lancaster, Virginia,

The Prayer at Valley Forge. With the fate of the infant nation hanging in the balance, Washington needed a miracle. He needed a huge delivery of fish. (Library of Congress)

and Wilmington, Delaware, the Commissary Department was hard-pressed to feed the army. Some of the troops were a day or more behind in rations by May 31. One officer wrote that "this day I believe we will be nearly able to furnish them with Shad. After that we must trust to Providence." He also reported that since June 4, 480 "bundles Cod Fish," at 56 pounds each, and 42 barrels of fish had been sent from Charlestown to Valley Forge. A brigade was a dozen or so wagons. These seem to have arrived, because from June 1 to June 7, about 18,000 pounds of fish were issued and the following week, about 42,300 pounds.

By the time General Washington finally broke camp at Valley Forge in June 1778, mass quantities of shad, herring, cod, and other fish were still being ordered and consumed by Washington's troops.

In this key moment of the American Revolution, fish helped win the day, since the Americans were saved from total annihilation. George Washington and his troops went on to lose most of the battles that followed over the next three years, but with the critical help of the French, they harassed, exhausted, outfoxed, and bled out the British so badly that they won the war.

Once the war was won, George Washington turned to his old pastime of recreational fishing as a way of relaxing and rejuvenating himself amid the stresses of building a new nation.

As President of the Federal Convention in Philadelphia of 1787, the pressures of trying to hammer out a new Constitution must have felt brutal to Washington.

So when the convention adjourned for two weeks off in the boiling summer of 1787, Washington did what any normal country boy would do—he went fishing. He even kept a diary of his angling adventures. His entry for July 30, 1787: "In company with Mr. Govr. [Gouverneur] Morris, and in his Phaeton with my horses; went up to one Jane Moores in the vicinity of Valley

Forge to get trout." For August 3: "In company with Mr. Robt. Morris and his Lady and Mr. Gouvr. Morris I went up to Trenton on another fishing party. Lodged at Colo. Sam Ogdens at the Trenton Works. In the evening fished, not very successfully." The next day: "In the morning, and between breakfast & dinner, fished again with more success (for perch) than yesterday."

In 1790, when the capital was New York City, now-President of the United States George Washington packed his tackle box and took a fishing trip over to nearby Sandy Hook, New Jersey, where he caught plenty of blackfish and striped bass.

Another time, to settle a policy dispute, Washington reportedly took rivals Thomas Jefferson, the Secretary of State, and Alexander Hamilton, the Secretary of the Treasury, on a fishing trip on the Delaware River. It's funny to try to picture those three distinguished Founding Fathers in their fishing outfits, whooping and hollering over a big fat fish they pulled out of the river!

According to the papers of Washington's private secretary Tobias Lear, Washington loved meals of salt codfish, and members of the New England Congressional delegation often kept him supplied with cod. One of his favorite Saturday dinners was boiled beets, potatoes, and onions mixed with the boiled codfish and covered with pork scraps and egg sauce.

Looking to the future of the United States, Washington thought of fish. When he picked the spot where the capital city should be on the Potomac, one of the reasons he liked it, he wrote, was that it was "at the head of a river plentiful with fish the year round."

Even the father of America was no stranger to two of the most familiar companions to the fisherman—frustration and disappointment.

In 1789, during a presidential grand tour of New England, Washington traveled up to Portsmouth, New Hampshire, where he decided to take out a little fishing boat. The entry in his diary read: "On Monday, November 2nd, 1789, having lines, we proceeded to the fishing banks a little without the harbor and fished for cod; but it not being proper time of the tide, we only caught two, with which, about 1 o'clock, we returned to town."

For almost his entire life, this greatest of Americans kept returning to fishing, through business crises, wars, political upheaval, and constitutional turmoil.

In doing so, George Washington proved the truth of an adage my dad told me a long time ago.

"Once a fisherman, always a fisherman!"

A new nation of vast potential was born, a fragile experiment in democracy that was often inspired by the highest forms of spiritual values, but was plagued by its "original sin" of human slavery, and vexed with the moral challenge of how to coexist with tens of millions of native people who had already achieved dominion over the continent for many thousands of years.

It was an infant nation that would rise and prosper, in part, on the backs of fish and other creatures of the sea.

3

The Greatest American Fishing Trip of All

There are always new places to go fishing. For any fisherman, there's always a new place, always a new horizon.

—Jack Nicklaus

Listen to the sound of the river and you will get a trout.

—Irish proverb

It was 3:30 P.M. on May 21, 1804.

Perhaps the most epic fishing trip in American history was about to begin.

At that moment, onlookers cheered as a core group of thirty-one men, many of them U.S. Army volunteers, pushed off from the banks of the Missouri River at St. Charles, Missouri, in a fifty-five-foot-long keelboat and two smaller boats.

Let's hop on board.

The men were headed west, into the unknown wilderness that lay beyond the American frontier. They were heading for some of the most spectacular fishing country on Planet Earth.

In charge of the expedition was U.S. Army captain Meriwether Lewis, the twenty-eight-year-old former personal secretary to President Thomas Jefferson, and his co-commander, thirty-three-year-old lieutenant William Clark.

They would live off the land and the water. This was a fishing trip inside a hunting trip, inside a voyage of discovery the likes of which America would not see again until the moon landing in 1969.

The boats were stuffed with enough gear to enable the voyagers to last two years on the edge of the known world. Naturally, they were toting plenty of good hunting gear, including a small arsenal of high-end rifles, muskets, powder, and shot to defend themselves and bag wild game for food.

They also packed a good supply of fishing tackle to work with—hundreds of fishing hooks, lots of fishing line, and a stave-reel to wind out the line on. They also packed 193 pounds of a portable dry soup mixture of veggies and beans, plus medicine, camping gear, a chronometer, presents and peace medals for the many Native Americans they expected to meet, scientific instruments, notebooks and ink, spare clothes, lead containers for the fragile items, and a letter of unlimited credit personally signed and guaranteed by President Jefferson himself.

Jefferson ordered the expedition, dubbed the "Corps of Discovery," to show the American flag through the newly acquired, mysterious lands of the Louisiana Purchase and northwest Columbia Basin, to find a smooth water route to the Pacific, and to make friends and launch trade with the Native Americans. And he wanted them to write it all down: the customs of the people they met, and the plants and animals and fish they saw. Fishing would not only provide an extra source of food for the Corps; it would be an emergency backup and an insurance policy for when the big game was too far away. And as it turned out, fish would save the entire expedition from probable starvation, not just once but several times.

In charge of the fishing was an expert angler by the name of Silas Goodrich, an Army private from Massachusetts who had a master's touch with hook and line. Lewis called him "our fine fisherman." Silas Goodrich was the kind of natural-born fisherman who knew where the fish were, how they moved, how they thought, how to catch them, and how to teach others to fish.

We don't know where he picked up these skills, but as soon as they hit the water, he was reeling in fish all over the place—eventually including shad, salmon, bass, pike, sauger, goldeye, trout, herring, and, the men's favorite, catfish, some as huge as 100 pounds. Goodrich and his buddies were probably the first white men ever to fish in the Rocky Mountains.

Sometimes they fished to add variety to their regular diet of buffalo, deer, elk, bear, pronghorn, birds, and small game like turkeys, rabbits, grouse, squirrels, and other game, or dog, which they bought for food from Native Americans in the Pacific Northwest. Other times, they literally fished for their lives. But always, they took animal life ethically and respectfully—they ate what they caught. Lewis set the tone when he wrote, "although game is very abundant and gentle, we only kill as much as is necessary."

In these days before fly reels were widely used in the New World, they used the old reliable hook-and-line method. It was the essence of simplicity—the hook was baited with grasshoppers or bits and pieces of deer or buffalo meat, you dropped the line into the water, you waited for a tug, gave a quick pull to catch the hook

U.S. Army Captain Meriwether Lewis, mastermind of the greatest fishing trip in American history. (Library of Congress)

in the fish's mouth, and you played it hand over hand to the surface. At different points of the journey, the expedition also caught fish with traps, big bush nets, or spears, which they called "gigs."

In the summer of 1804, as the expedition floated west on the Missouri River past the grasslands of the Great Plains and present-day Missouri and Nebraska toward South Dakota, the fishing got intense.

They were catching dozens of catfish per day, each averaging 30–40 pounds. During a stopover near present-day Salix, Iowa, they hit the jackpot. On

Lieutenant William Clark, co-commander of the expedition. (Library of Congress)

August 15, 1804, Clark wrote, "I went with ten men to a creek dammed by the beavers"; "with some small willow & bark we made a drag and hauled [it] up the creek, and caught 318 fish of different kind, pike, bass, salmon, perch, red horse." Three hundred and eighteen fish in one day—that's one incredible day of fishing!

But the next day was even more spectacular. Captain Lewis took twelve men out fishing, and they came back with 800 fish, including 490 catfish, which they were delighted by. Catfish are the original all-American, rugged, bottom-feeding fish, common to ponds and

Catfish. (New York Public Library Digital Collections)

rivers across much of the United States. If you cook them up right like we do down in Louisiana—filleted and rolled in cornmeal, gently fried till golden brown, flavored with a smack of cayenne pepper and Cajun spice—they're as tasty as fish can get. For Southerners, they're the ultimate meat-and-potatoes, down-home, backcountry fish, and today they're the star of a multibillion-dollar American commercial farming business that employs tens of thousands of people. Sports fishermen love them, too. "There's just something about casting for catfish after dark that defines the perfect summer night. It could be the campfire camaraderie, but more likely it's the thrill to be had not knowing what kind of monster is sniffing your bait," wrote the editors of *Field & Stream* in 2013. "Catfish are warmwater fish, and by midsummer they're feeding heavily to gain back the weight they lost during the spawn. Prime time is between 10 p.m. and 2 a.m. when fish move out of the deep to feed. No expensive lures here—just a simple Carolina rig with an egg sinker above a 3/0 circle hook tipped with cutbait, shrimp, or a homemade blood bait for the more adventurous. Once you've cast this out into the dark, sit back and steel yourself—for you never know what may go splash in the night."

While catfish aren't exactly beautiful, they are fascinating critters, with eight signature "whiskers" or "barbels" that are actually highly sensitive taste and smell organs for locating food. Some catfish even have a natural "communications device" built into their pectoral fins that sends out drumming-like patterns that may be connected to courtship and conflict rituals. How big can these underwater varmints get, say a giant blue catfish? How about 5 feet long and 200-pounds plus! That puts them up near WWE wrestling range. Some of their monstermouths are so huge they can scarf down a basketball.

A few years back, I learned of a great way to catch catfish. Fill up a burlap bag with a dozen chicken livers, the bloodier the better. Put the bag inside a sturdy plastic bag. Mash out the livers thoroughly with a rolling pin. Let the whole deal "age" in the sun for a few days, until it gets extra funky. Tie the plastic bag extra tight to assert some control on the stink. Find a nearby catfish hole, drop a rock inside the bag, and let the whole thing sink to bottom, where it will ooze off delicious scents that will pull in catfish. After a couple of hours, bait a No. 2 Eagle Claw hook with a hunk of liver and lower it down near the burlap bag. You may be able to pull in a whole barrel full of catfish this way.

Out in the wilderness in the summer of 1804, Lewis, Clark, and their teammates savored delicious meals of catfish fillet and wild berries as they pushed westward. On September 1, in present-day South Dakota, Clark wrote of "numbers of catfish caught, those fish is so plenty that we catch them at any time and place in the river."

The Corps of Discovery spent the winter of 1804–05 at a spot near present-day Washburn, North Dakota, they dubbed Fort Mandan. On April 7, 1805, they sent several temporary expedition members and the keelboat back to St. Louis, along with a trove of maps, reports, and specimens Lewis and Clark had prepared so far. Joining the Corps was a Shoshone Native American interpreter named Toussaint Charbonneau, his mate, a teenage girl named Sacagawea, and her infant boy, Jean

Baptiste. Sacagawea was a great help to the expedition all around, and she pitched in with the fishing, too.

Not much fishing happened in the first weeks after the Corps set off in eight canoes westward into Montana in the spring of 1805. They didn't often feel the need to fish, since they were in the midst of vast herds of buffalo, their favorite food, plus elk, deer, and antelope. From here until reaching the West Coast, they were traveling in lands that few if any white men had ever seen before.

On June 13, 1805, during a scouting mission that found the Great Falls of the Missouri River, Lewis wrote of a discovery by his chief fisherman: "Goodrich had caught half a dozen very fine trout and a number of both species of the white fish, these trout are from sixteen to twenty three inches in length, precisely resemble our mountain or speckled trout in form and the position of their fins, but the specks on these are of a deep black instead of the red or gold color of those common to the United States, these are furnished long sharp teeth on the pallet and tongue and have generally a small dash of red on each side behind the front ventral fins; the flesh is of a pale yellowish red, or when in good order, of a rose red."

This was the first known record of non–Native Americans ever seeing a cutthroat trout, a striking-looking fish with a signature streak of bright crimson under its jaw. In honor of the expedition, the species was later given the scientific name of *Oncorhynchus clarkii*, or *Salmo clarkii*, after the expedition's co-commander. To me, the cutthroat is a magical kind of creature, like its cousins the native-to-America rainbow (also called steelhead when it can transition from salt water to freshwater), the brook trout, and the brown trout, which was transplanted to America from Europe later in the 1800s. Today, trout are widely distributed around the United States, clear up to Alaska, where you'll find some of the biggest rainbow trout.

Trout are great fun to fish, delicious to eat, and downright beautiful to behold. "There are many kinds of natural beauty, from the expansiveness of a sky showing through a covering of multigreen forest, to the myriad forms of wondrous natural life," marveled outdoor writer Mike Hudoba. "But there is a special something, a washing of the spirit, a re-creation of the individual, for those who take to the streamside where trout waters sing." Like many other fish species such as bass, northern pike, and sunfish, trout spend a lot of time near the shade, shelter, and protective cover of underwater "structure": downed trees, stump fields, weeds, islands, brush, rock piles, grass beds, and drop-offs. A rainbow trout can live up to seven years, and a lake trout can live

Rainbow Trout. (New York Public Library Digital Collections)

well over twenty years. Trout are partial to cool lakes and streams, and like to eat juicy grasshoppers, night crawlers, cheese, bread, mayflies, minnows, crayfish, and other fish. An experienced angler can "play" a big trout by putting light finger pressure on outgoing extra line, then using the reel's drag system to tucker out the fish and bring it in. Once caught, the delicate trout must be handled very carefully before releasing.

But Lewis and Clark expedition chief angler Silas Goodrich wasn't practicing "catch and release"; this was all straight-up "catch and eat" fishing, and soon he was pulling up specimens of two more species brand-new to the white man: sauger and goldeye. Lewis and Clark carefully studied the fish and filled their journal books with detailed observations.

In the days that followed, as the Corps camped around the Great Falls, Captain Lewis tried his luck at fishing by baiting his hook with "melt," or deer spleen. On June 19, he wrote, "I amused myself in fishing several hours today and caught a number of both species of the white fish, but no trout nor cat."

In July, as the expedition floated upstream through the valleys of southern Montana's soaring Rocky Mountains, fish were plentiful, but became much harder to catch. "We see a great abundance of fish in the stream some of which we take to be trout," wrote an annoyed Lewis on July 29, "but they will not bite at any bait we can offer them." Good news came later that day, when Clark reported that the Corps "caught three large catfish," and that "those fish are in great plenty on the sides of the river and very fat, a quart of oil came out of the surplus fat of one of these fish."

But soon both big game and fish were thinning out. Lewis began worrying that they might run out of food. "I know when we leave the buffalo country that we shall sometimes be under the necessity of fasting occasionally," he wrote on July 3, 1805.

In August, the Corps reached the navigable limit of the Missouri River and turned into the Jefferson River, heading south. They crossed the Continental Divide at Lemhi Pass in present-day Idaho, and suddenly entered one of the richest salmon fisheries on earth. Every spring and fall, sockeye, coho, and chinook salmon swarmed by the millions into the streams and rivers of the Pacific Northwest. From here until they hit the Pacific Ocean, salmon would often be the expedition's salvation.

Lewis and Clark had entered the American Empire of Pacific Salmon, a dominion that stretched over the western part of the continent from present-day California, clear up to Alaska. Many tribes in the Pacific Northwest depended on tasty, protein-rich, hard-fighting salmon for their survival, and most of the major rivers in the region were interlaced with Native American salmon-fishing weirs, dip nets, poles, platforms, and settlements. The biggest salmon, the Chinook salmon (also known as king salmon or blackmouth salmon), often exceeds 30 pounds, and the female Chinook can lay over ten thousand eggs. Salmon are long-distance, dual saltwater and freshwater travelers—on their spawning run from the Pacific Ocean to central Idaho, for example, sockeye and Chinook salmon from central Idaho climb almost seven thousand feet, and journey more than nine hundred miles. In the centuries that followed, both Pacific and Atlantic salmon became, like their close relatives the trout, one of America's most beloved fish, both for sportfishing and eating.

Salmo Salar, The Salmon. (New York Public Library Digital Collections)

On August 13, 1804, Meriwether Lewis savored his first piece of salmon. He noted in his journal, "on my return to my lodge an Indian called me in to his bower and gave me a small morsel of the flesh of an antelope boiled, and a piece of a fresh salmon roasted; both which I eat with a very good relish, this was the first salmon I had seen and perfectly convinced me that we were on the waters of the Pacific Ocean."

The presence of salmon also tipped Lewis and Clark off that they were within range of their ultimate destination, the Pacific, although one look at the steep walls of the Bitterroot Mountains revealed that there would be no all-water route to the ocean. They would have to push westward by horseback and canoe with the help of local Native Americans. Over the next weeks, the Corps traded gifts of fish and game with local Shoshone Native Americans, a mountain valley tribe whose main food was fish.

On August 21, 1805, the Native Americans showed Clark their elaborate fish trap on the Lemhi River, an intricate network of log dams, stakes, poles, and twenty-foot-long cylindrical baskets. They also put on a demonstration of salmon spearing, or "gigging," for Clark, who was fascinated by their technique. "Their method of taking fish with a gig or bone is with a long pole, about a foot from one end is a strong string attached to the pole," Clark wrote. The Native Americans "turn and strike them so hard that the hone passes through and catches on the opposite side, slips off the end of the pole and holds the center of the bone." The Native Americans dried enough fish to last through the winter.

The next day, inspired by the Native Americans' technique, the Corps put together their own fish trap, and hit the jackpot. "Late in the evening," wrote Lewis, "I made the men form a bush drag, and with it in about 2 hours they caught 528 very good fish, most of them large trout, among them I now for the first time saw ten or a dozen of a white species of trout, they are of a silvery color except on the back and head, where they are of a bluish cast, the scales are much larger than the speckled trout." He added, "I distributed much the greater portion of the fish among the Indians." The "white trout" Lewis described may have been the steelhead, a light-silver–tinted rainbow trout, or a west slope cutthroat.

Any experienced fisherman will tell you that fishing can be an emotional roller coaster, and that's exactly what it felt like to Lewis and Clark in the Rocky Mountains. Just a few hours after distributing hundreds of fish to their Native American friends, the Corps fished and fished, but kept coming up empty. The next night, Lewis, now on the North Fork Salmon River near present-day Salmon, Idaho, lamented, "we attempted to gig fish but were unsuccessful, only obtaining one small salmon."

On August 27, things looked especially bleak, and some of the men feared they would starve. Clark wrote, "one man killed a small salmon, and the Indians gave me another which afforded us a slight breakfast." He noticed that the stoic local Native Americans were content to survive on the fish they could catch, but his own men, by contrast, were "hourly complaining of their retched situation and [having] doubts of starving in a country where no game of any kind except a few fish can be found."

Native American fisherman working the Columbia River, near the final destination of Lewis and Clark's Corps of Discovery mission. Lewis and Clark were repeatedly rescued by Native Americans—and by fish. (Edward S. Curtis/Library of Congress)

That night, Clark reported, "an Indian brought into the camp five salmon, two of which I purchased which afforded us a supper." This pattern, of the Corps being rescued by Native Americans and/or fish, would be repeated many times during the voyage.

On August 28, Clark complained that the salmon the Corps were eating were tasty, but they somehow sapped him of energy. "Those salmon which I live on at present are pleasant eating," he wrote, but "they weaken me very fast and my flesh I find is declining." It might not have been the salmon's fault, since the men were exerting themselves terribly in the rain, cold mountain streams, and rugged horseback rides, dragging huge amounts of provisions. They were eating little else in the way of protein, vegetables, sugar, or grains, and were probably starving for many different nutrients.

In late September and early October, as they traveled through Nez Percé territory, the expedition survived mainly on dried salmon and Camas roots sold to them by the Native Americans, most of whom were being amazingly hospitable to the travelers.

The good news was the food kept them alive, as there were no buffalo (their favorite trail food) in the high mountains, and deer were usually nowhere to be seen in the dense forest, just the occasional squirrel or grouse.

The bad news was the salmon-and-roots diet, which, combined with physical exertion, made them wickedly sick. "I find myself very unwell all the evening from eating the fish and roots too freely," Clark wrote in his journal. By September 24, Captain Lewis and eight of his men were ill. Clark wrote three days later, "nearly all the men sick," complaining of "a lax and heaviness at the stomach."

A week later they had recovered enough to leave their horses with the Nez Percé Native Americans, load their canoes, and head down for the Pacific via the Clearwater and Snake Rivers, and the Columbia River, the largest river of the American West, which flows 1,200 miles from southeastern British Columbia to the Pacific.

On October 24, near the present-day town of The Dalles, Oregon, the expedition passed a fishing station of the Chinook tribe that featured no fewer than one hundred drying racks for fish. On the twenty-sixth, Clark described an especially scrumptious fish dinner: "one man gigged a salmon trout which we had fried in a little bears oil which a chief gave us yesterday and I think the finest fish I ever tasted."

On November 7, the expedition achieved its destiny, when the Corps spotted the estuary of the Columbia River. "Great joy in camp we are in View of the Ocean," wrote Clark. Now all they had to do was camp out and hunker down through a cold, rainy Pacific Northwest winter, somehow stay alive until spring, and then climb, ride, and float back east all the way they came, to deliver their maps, journals, and specimens to President Jefferson.

By now, the explorers were completely exhausted. And they were totally sick of fish. They were desperate to find some elk, deer, or other larger game to sink their teeth into, but their hunting teams came up empty. They wandered around the Pacific Coast, trying to figure out where to set up camp. On November 11 they met a group of local Clatsop Native Americans and traded some fishing hooks for a dozen sockeye salmon. They next day, drenched from constant rain, they speared sixteen salmon on the river.

On November 28, after the hunters again found no deer in the forest, Clark practically cried into his journal. "We have nothing to eat but a little pounded fish which we purchased at the Great Falls. This is our present situation! Truly disagreeable, added to this the robes of ourselves and men are all rotten from being continually wet. O how tremendous is the day. This dreadful wind and rain continued with intervals of fair weather, the greater part of the evening and night."

Four drizzly days later, the Corps was dreaming of elk, and couldn't find any. The Corps members were practically within spitting distance of the greatest salmon fisheries on earth, but for some reason they refused to go out and catch enough fish themselves to tide them over. Maybe they were just too wet, cold, and sick of salmon to do it. "I am very unwell," Clark wrote; "the dried fish which is my only diet does not agree with me and several of the men complain of a lax, and weakness."

On December 9, while exploring the hills, Clark met three Clatsop Native Americans who were loaded down with fresh-speared salmon. They invited Clark to their village, where he was given gifts of salmon, root veggies, berries, and a tasty berry syrup.

The next day, Clark wrote, "I saw Indians walking up and down the beach which I did not at first understand the cause of, one man came to where I was and told me that he was in search of fish which is frequently thrown up on shore and left by the tide, and told me the 'sturgeon was very good' and that the water when it retired left fish which they ate." It seemed that the Native Americans made it through the winter largely by eating fish they found tossed up on the beach.

That month, the voyagers mustered enough energy to build Fort Clatsop on Columbia River, and a satellite beach camp on the Pacific shore fifteen miles away. The local Native Americans from another nearby tribe, the Tillamook, treated the Corps of Discovery team to a special feast of root vegetables and boiled blubber from a 105-foot-long whale that they had harvested. On January 8, Clark wrote, "I prize it highly; and thank Providence for directing the whale to us; and think him much more kind to us than he was to Jonah, having sent this monster to be swallowed by us instead of swallowing of us as Jonah's did."

On January 16 and 17, Lewis made careful notes of Native American fishing techniques during the peak spring and summer seasons. Using fishing lines and nets made of silk-grass or white cedar bark, sharp-angle hooks fashioned from bones, and spears, the Native Americans went after salmon and other species in marshlands and islets around the mouth of the Columbia River, including the hard-fighting white sturgeon, which can grow up to 13 feet and can weigh over 700 pounds.

A hopeful sign appeared on February 13, wrote Lewis, when "the Indians inform us that we shall have great abundance of a small fish in March which from their description must be the herring." But until then they had to hang on through the freezing, wet winter, scrounge for fish, and hope the Native Americans would be kind to them.

By February 14, Clark had finished his greatest accomplishment, a detailed map of the new lands of the United States from Fort Mandan to the Pacific Coast. By then Lewis and Clark concluded that they had fulfilled their mission. They had established a path across the continental United States. It wasn't an all-water route, since they had to switch to horses and to different boats at times, but they had blazed the trail for President Jefferson and America.

In late February, nearly all the travelers were immobilized by colds, flu, and fever, and were in no shape to go on strenuous forays for game and fish. On February 24, fish-bearing Native Americans again came to the rescue, wrote Clark, to sell "sturgeon and a species of small fish which now begin to run [eulachon, sometimes called candlefish], and are taken in great quantities in the Columbia River about 40 miles above us by means of skimming or scooping nets."

A grateful Lewis was delighted by the flavor of the eulachon, of which he wrote, "I think them superior to any fish I ever tasted." His partner Clark agreed, calling them "deliciously fine." In a few days, when the eulachon ran out, a hungry Lewis was relieved when one of the men "arrived with a most acceptable supply of fat sturgeon, fresh anchovies and a bag containing about a bushel of wappetoe [a root vegetable]." They feasted sumptuously on the haul, and took careful note of how the Native Americans used techniques like pickling, smoking, and steaming their fish.

In March, Lewis wrote of how the Native Americans loved to eat starry flounder, and harbor porpoise, the taste of which he hated. The Chinook salmon and sockeye salmon, he noted, were so plentiful in the sea and rivers that the "natives are so much indebted for their subsistence." In mid-March, Lewis made journal notes about clams, mussles, periwinkles, cockles, and shellfish. A Native American brought him a freshly caught white coho salmon, which signaled the coming of the great springtime salmon runs on the Columbia.

But on March 5, the Corps was again running out of food, and land animals that could sustain them were nowhere to be found. Lewis wrote that once again, hunters were sent out but came up empty: "They had neither killed nor seen any Elk. They informed us that the Elk had all gone off to the mountains a considerable distance from us. This is unwelcome information and rather alarming. We have only two days provision on hand, and that nearly spoiled."

The next day, fortunately, a Native American appeared in camp with a few eulachon, which helped ward off starvation, and Lewis sent Sergeant Pryor up the Columbia with some trading items. His mission: find Native American fishermen, and bring back fish. The men hung on until March 11, when rescue came: "Early this morning Sgt. Pryor arrived with a small canoe loaded with fish which he had obtained from the Cathlahmah's [Indians] for a very small part of the articles he had taken with him."

They ate the fish, and again ran through the supply. Then it hit him. Lewis realized that the Corps of Discovery could expect little or no time to hunt wild game to sustain them on the first part of their long return journey back to St. Louis. If they couldn't get their hands on some decent-sized fish, lots of them, to fuel them on the strenuous paths eastward through the mountains, they could die.

Luckily, help was on the way, wrote Lewis, as "the Indians tell us that the Salmon begin to run early in the next month; it will be unfortunate for us if they do not, for they must form our principal dependance for food in ascending the Columbia above the Falls and its S. E. branch."

On March 23, 1806, after almost two fish-fueled years in the wilderness, Lewis and Clark headed for home.

Along their way back up the Columbia River, they had good luck fishing, and they kept on buying dried fish from local Native Americans to supplement their catch. On April 19, 1806, at Celilo Falls, near the present-day border of Oregon and Washington state, Lewis and Clark apparently witnessed an event of deep spiritual meaning to the native peoples of the Pacific Northwest. It was a Native American "First Salmon Ceremony." Clark wrote, "There was great joy with the natives last night, in consequence of the arrival of the salmon. One of those fish was caught. This was a harbinger of good news to them."

The ceremony, which was common to native peoples throughout the Columbia River basin, differed from tribe to tribe, but according to historian John Harrison had several things in common. "The salmon chief of the tribe would select a fisher to catch the first salmon," wrote Harrison. "This was an honor, and before entering the river the fisher would undergo a blessing or a purification. Once a fish was caught, it would be brought to shore and carefully prepared, cooked and distributed to the people in a manner unique to the location and tribe. The head of the fish would be kept pointed upriver to show the salmon's spirit the way home. The bones would be carefully cleaned and returned to the river, where it was believed the salmon would reconstitute itself and continue its journey."

Carla HighEagle, a leader of the contemporary Nez Percé tribe, explained, "The salmon was put here by the Creator for our use as part of the cycle of life. It gave to us, and we, in turn, gave back to it through our ceremonies. . . . Their returning meant our continuance was assured because the salmon gave up their lives for us. In turn, when we die and go back to the earth, we are providing that nourishment back to the soil, back to the riverbeds, and back into that cycle of life."

When Lewis and Clark's Corps of Discovery finally arrived in St. Louis on September 23, 1806, they became national celebrities. Their achievements in exploring the great Missouri and Columbia river systems were indeed breathtaking. They had traveled more than eight thousand miles in under two and a half years and forged a path across the continent that inspired a huge boom in trade and settlement. They established friendly contact with over twenty native tribes, despite an unfortunate shoot-out with Blackfeet Native Americans in which two natives were killed. They were the first to make scientific descriptions of several new animal species, including the mountain goat, pronghorn antelope, prairie dog, and grizzly bear. And they described a number of new species of fish, including Yellowstone trout, cutthroat trout and West Slope trout, blue catfish, channel catfish, sauger, and starry flounder.

Without the help of Native American Fishermen, and a fisherman named Silas Goodrich, and a few thousand juicy salmon and other fish, Lewis and Clark never would have made it.

Silas Goodrich, unfortunately, fell victim to syphilis, a disease that Clark reported the fisherman "contracted from an amorous contact with a Chinnook damsel" during the expedition. He should have gone fishing instead. He was dead by the early 1820s.

Meriwether Lewis started drinking heavily at the parties that welcomed the heroes home, and he never stopped. He died at age thirty-five from a mysterious gunshot wound.

As for co-captain William Clark, well, he went on to live a clean, productive life and died at the then ripe old age of sixty-eight.

After his epic voyage, we don't know if he ever ate salmon again.

The fish-powered Lewis and Clark expedition marked the symbolic opening up of the American West. It showed how bountiful the West was, and how much potential and opportunity lay beyond the horizon. It marked the birth of the United States as a continental power, as millions of dreamers, traders, cowboys, hunters, fishermen, and settlers soon followed the path west. They were propelled by noble ideals, commercial interests, and for some, a sense of "Manifest Destiny" of America as a world power and moral exemplar. They brought with them waves of prosperity and human progress—and intensifying struggles over slavery and Native American rights.

Barely three generations after Thomas Jefferson launched the Corps of Discovery's epic adventure, the United States faced its most dangerous challenge, a conflict that began to literally tear the nation apart. And at a key moment in climax of the drama, fish played a pivotal role.

4
The Fish That Won the Last Battle of the Civil War

This means victory. This *IS* victory!

—Abraham Lincoln

Did you know that a few dozen fish helped decide the last battle of the Civil War?

It's a true story, and it happened on April 1, 1865, at a place called Five Forks, Virginia.

At that spot, 17,000 Union infantry and cavalry troops led by General Philip "Little Phil" Sheridan approached from the southeast toward 10,000 Confederate troops led by Major General George E. Pickett (who achieved fame for the failed "Pickett's Charge" at the Battle of Gettysburg), dug in along two miles of ditches and fences. The rebel troops were guarding the critical

South Side Railroad supply line that linked to General Robert E. Lee's bigger force guarding the besieged city of Petersburg. If the rail line was cut, Lee would have to abandon Petersburg, the federal troops could capture the Confederate capital of Richmond, and the rebel cause would be lost.

Around midday, General Pickett received some delightful news. A Confederate division commander named Thomas L. Rosser invited him over for an early dinner of locally caught, fresh-baked fish. What better way to ingratiate himself with his boss? The fish were shad, the same species that George Washington caught at Mount Vernon and nourished his troops with at Valley Forge.

Rosser's troops had just caught enough shad from the first spring spawning run on the Nottoway River to feed a small group of officers, and Pickett jumped at the opportunity to indulge in anything other than the uniformly lousy Confederate Army chow he had endured for the last four years of battle. Pickett rounded up a few of his aides plus Major General Fitzhugh Lee (nephew of Robert E. Lee), and galloped off to the shad bake north of Hatcher's Run, at a spot a mile and a half north from Five Forks. Shad had a special place in many an American fisherman's heart. They were a favorite meal for Native Americans, and were later

Confederate Major General George E. Pickett, whose hunger for fish triggered the collapse of rebel forces at the biggest battle of the climax of the Civil War. (National Archives)

served by President John F. Kennedy at the White House.

Though bony and hard to cook correctly, there was something lyrical, poetic, and even heroic about shad. After all, the spring shad spawning run rescued many colonial and frontier settlers from death by starvation after brutal American winters.

They looked beautiful, too. Author John McPhee once wrote of a school of shad descending the Delaware River at twilight, churning the water white: "They dimple, dapple, leap into the air. They seem like a squall of silver rain. It has long been said that they are jumping for insects. What the human eye sees, as it observes the dimpling of the shad on the blue river bordered with turning leaves, is early Impressionism rooted in the Hudson River School—the peace and quiet of Nature with touches of silvery motion in it, an Arcadian pastoral vision." In the nineteenth century, shad had become such a staple meal that *The Philadelphia Record* noted in 1904, "Of all the food fishes of America, shad may be regarded as the most popular; it appears on the table of the wealthy epicure as a delicacy; while to the unpampered palate of the working man it forms a substantial and nourishing dish."

Incredibly, at Five Forks, with visions of baked shad dancing in their brains, both Generals Pickett and Lee neglected to tell their troops they were skipping off for the meal, where they spent much of the fateful afternoon of April 1, 1865.

Soon after the Confederate generals disappeared to the rear, Phil Sheridan's Union troops and cavalry

Shad. (New York Public Library Digital Collections)

launched a series of full-scale attacks that punched straight into the Confederate lines, including an attack led by a golden-haired swashbuckler and brigadier general by the name of George Armstrong Custer.

As they savored their fish meal, Pickett and Lee couldn't hear the commotion because it was muffled by an "acoustic shadow" created by thick woods and heavy, humid atmosphere. In fact, the "food was abundant" at the shad bake, wrote historian Douglas Southall Freeman, and "the affair was leisured and deliberate as every feast should be."

As Civil War historian Shelby Foote explained, "Neither told any subordinate where he was going or why, perhaps to keep from dividing the succulent fish too many ways; with the result that when the attack

Abraham Lincoln, photographed February 5, 1865, by Alexander Gardner. When Lincoln was shown proof of the Confederate's fish-triggered debacle at the Battle of Five Forks, he was jubilant. He knew the war was finally won. (Library of Congress)

The Battle of Five Forks: General Philip Sheridan pierces rebel lines at the critical moment of the battle. This moment triggered the end of the Civil War. It happened because of a fatal Confederate error involving baked fish. (New York Library Digital Collections)

exploded—damped from their hearing as it was by a heavy stand of pines along Hatcher's Run—no one knew where to find them. Pickett only made it back to his division after half its men had been shot or captured, a sad last act for a man who gave his name to the most famous charge in a war [Pickett's Charge] whose end was hastened by his three hour absence at a shad bake."

His belly full of fish and liquor, Pickett didn't get back to his troops in time. He was three hours late.

Close-up view of General Sheridan at the key moment. (New York Public Library Digital Collections)

"At that point," reported a witness, "the Confederate forces had been completely overrun." That's a nice way to put it. They were annihilated. Up to four thousand rebel prisoners were taken and Pickett narrowly escaped capture himself as he retreated in a panic. According to historian Joseph Wheelan, "Five Forks was the Army of the Potomac's first important battlefield victory since Gettysburg and arguably the most consequential of the war. By destroying Pickett's task force, the Cavalry Corps and V Corps had made Petersburg and Richmond untenable for Lee's besieged army. Lee could not stop Grant's men from capturing the Southside Railroad and smashing his right flank—or, for very long, from capturing the Confederate capital. The Confederacy's downfall was imminent." A single fish dinner spelled final doom for the Confederate States of America, at the last big battle of the Civil War.

"This has been the most momentous day of the war so far, I think," wrote Colonel Charles Wainwright, a Union artillery commander, "a glorious day, a day of real victory."

When President Abraham Lincoln was presented with a pile of battle flags captured at the battle, he was jubilant. He said, "Here is something material—something I can see, feel and understand. This means victory. This *is* victory!"

Eight days after the shad bake at Hatcher's Run, on April 9, 1865, General Robert E. Lee surrendered the Army of Northern Virginia to General Ulysses Grant at Appomattox Court House, Virginia.

The Battle of Five Forks was the "Waterloo" of the Confederate States of America. And it happened because of a little bunch of tasty fish.

The United States was whole again, a nation reborn, and while millions of black, native, and other Americans would face decades of tremendous struggle, the nation was now free to rebuild, expand, invent, innovate, pray, flourish—and fish.

Pickett's disaster at the Battle of Five Forks wasn't the only time the love of fish triggered a colossal American military screwup.

An equally historic episode happened in June 1876 at another "inflection point" in our history, during the last big battle of the wars between the U.S. Army and the forces of the Native American nations. It occurred in Montana Territory, at the climax of the Great Sioux War, when the colorful George Armstrong Custer, now back to his pre–Civil War rank of lieutenant colonel, moved in to destroy a multi-tribal force of native warriors under the command of Sitting Bull at Little Bighorn. Or so he thought.

That summer, the U.S. Army sent three columns of troops to crush the Native American confederation from two directions. One was commanded by General Alfred H. Terry, the mission's overall commander, and included five companies of the elite Seventh Cavalry led by Custer. This force and a column led by Colonel John Gibbon would head south and west. The third column, composed of some 1,200 soldiers marching north from Fort Fetterman in Wyoming, was led by Brigadier General George Crook, an avid outdoorsman.

General Crook, as it happened, was a hard-core fisherman and hunter. He traveled with a pack of his own private fishing gear, which included several split-bamboo fly rods, and kept a sharp eye out for lucky fishing spots. This simple fact may have spelled doom for General Custer.

With pristine, unspoiled waters and game ranges, this was a fantastic fishing region, not far from where Lewis and Clark fished when their epic journey passed through the region seven decades earlier. General Crook was spellbound by its beauty and its wildlife. The area was surrounded by rugged mountains, alpine meadows, desolate moonscapes, and grasslands rich with bison, mule deer, antelope, prairie dogs, and bighorn sheep. "Crook made the most of the terrain," wrote historian John H. Monnett, "often riding several hundred yards out in front of his troops in hostile Indian country, on the off-chance that he might flush out a sage hen or sharp-tailed grouse."

On June 17, 1876, near the Yellowstone River, Crook's column blundered into the operation's first contact with the enemy, in an encounter called the Battle of the Rosebud River. A large Sioux force commanded by legendary fighter Crazy Horse surprised the Army troops and blocked their advance about twenty miles from Little Bighorn. Casualties on the Army side were 28 killed and 56 wounded. If Crook had kept moving north after the encounter, he might have pushed the Native American fighters into Terry's formation before Custer got separated from it days later, and he might have arrived at Little Bighorn in time to reinforce Custer.

But the ordinarily skilled and aggressive military officer Crook abruptly threw in the towel and skedaddled. His force was low on ammunition and tired from the fight, and Crook pulled back to a comfortable spot at Goose Creek, set up camp, sent back his wounded, and refused to budge from the area until reinforcements came.

Then he broke out his fishing tackle.

On the receiving end of Crook's request for backup troops was General Philip Sheridan, now commander of the Military Division of the Missouri, the same Phil Sheridan who successfully exploited Confederate General George Pickett's ill-timed fishing trip at the Battle of Five Forks the previous decade. Now it was a furious Sheridan's turn to be clobbered by a fishing catastrophe.

Brigadier General George Crook, U.S. Army. He could have reinforced George Custer at Little Bighorn. Instead, he and his men launched an epic assault on the fish population of Montana Territory, abandoning Custer and his men to their doom. (National Archives)

What did General Crook do at Goose Creek? Well, for over a solid week, he and his men went on an orgy. A fishing orgy. "The streams around his base camp near the headwaters of the Tongue River were choked with trout," wrote historian Jim Merritt, "and the country teemed with elk, antelope, and mountain sheep."

Let's go fishing, boys! was the common, irresistible urge among the troops.

What happened next would be hard to believe if a Captain John Gregory Bourke hadn't written it all down. Bourke was both a witness and a participant in the fishing rampage. Using artificial flies and natural bait like live grasshoppers, according to Bourke, the soldiers began snagging huge numbers of cutthroat trout.

At first, the goal of the troops was to each catch up to thirty trout per day. But then, Bourke explained, many of them got "carried away by the desire to make a record [catch] as against that of other fishers of repute." Soon some soldiers were catching eighty trout in a single day. Before long, "hundreds and thousands of fine fish [were] taken from that set of creeks by officers and soldiers," thanks to the "wonderful resources of the country."

The commander set the tone, of being both a responsible and ethical fisherman and a fishing champion. "General Crook and the battalion commanders were determined that there should be no waste, and insisted upon the fish being eaten at once or dried for later use." Crook reeled in seventy fish in a single day, inspiring a team of three enlisted men to double his score in one afternoon by hooking 146 trout.

One day, Captain Bourke peeled off with some buddies to investigate the action in some alpine lakes, but found the cutthroat to be especially cagey. "We tried them with all sorts of imported and manufactured flies of gaudy tints or sombre hues, [but] it made no difference," he recalled. "After suspiciously nosing them they would flap their tails, strike with the side-fins, and then, having gained a distance of ten feet, would most provokingly stay there and watch us from under the shelter of slippery rocks. Foreign luxuries evidently had no charm for them."

The officer had better luck with a cutthroat at a nearby stream. "I gave him all the line he wanted fearing I should lose him. His course took him close to the bank, and, as he neared the edge of the stream, I laid him, with a quick, firm jerk, sprawling on the moss. I was glad not to have had any fight with him, because he would surely have broken away amid the rocks and branches. He was pretty to look upon, weighed three pounds, and was the largest specimen reaching camp that week. He graced our dinner, served up, roasted and stuffed, in our cook Phillip's best style."

The soldiers seemed lost in a collective fishing trance. General Crook's Great Fishing Party continued on and on, as the men of Lieutenant Colonel Custer's now-detached 260-man force faced their doom. On June 25, 1876, after a series of bad tactical decisions by Custer, he and his men were annihilated by an overwhelming combined force of Sioux and Cheyenne warriors. It was the last big battle of the Indian Wars, and the United States lost. As Custer's force was being destroyed, General Crook and his over thousand-man formation were nowhere to be seen. Three days later, not far away, General George Crook and his men were still in a fishing frenzy, scoring a banner five hundred trout in twenty-four hours.

Custer's defeat was "a spectacular triumph for the American Indian in his four-century struggle to hold back the white people who finally overpowered him," in the words of Robert Utley, the chief historian of the U.S. National Park Service. News of the disaster shocked the nation on the eve of its centennial celebration, but the Native American victory was short-lived. The U.S. government intensified its campaign to subdue native resistance, and within a decade, all the major tribes had abandoned armed resistance and moved to dirt-poor conditions in Native American reservations, where many of their descendants reside today.

Not long before he made his fatal "Last Stand" at Little Bighorn, Custer had inexplicably divided his forces and sent three of his five companies of cavalry troops on a futile maneuver around the right flank of the Native American encampment.

One of the companies was led by a man whom Custer hated, Captain Frederick Benteen, a courageous veteran of the Civil War and a native Virginian who stayed loyal to the Union cause. Custer and Benteen were like hateful spouses trapped in a bad marriage, constantly sniping at each other behind their backs. Some historians figure Custer wanted Benteen out of the way so he could grab all the glory of victory. Maybe Custer was so sick of his cantankerous subordinate that he just preferred him out of sight for a while. Whatever the reason, it was a fatal error. Custer needed all the firepower he could get right beside him, as the native force was much bigger that he expected.

Captain Benteen returned to the scene of the Last Stand too late (some later accused him of dragging his feet and thus dooming Custer), just as the last of Custer's men were being chewed to pieces by enemy gunfire, arrows, and spears. But over the next two

Captain Frederick Benteen, who could not rescue Custer, but achieved immortality by leading a charge while waving his fishing rod in the air. (Smithsonian)

days, Benteen rallied the remnants of Custer's larger force into a perimeter defense and fighting retreat that saved their lives.

Benteen had one more moment of glory, the year after the Battle of Little Bighorn—as a fisherman. Like General Crook and a great many military officers at the time, Benteen was a devoted angler, and one of his troopers remembered, "I saw him wade over his boot tops many times into cold water to get mountain trout."

On September 13, 1877, at the Battle of Canyon Creek, Captain Benteen led a wild Seventh Cavalry charge against the Nez Percé encampment on the Yellowstone River. As he galloped toward the enemy, bullets flying all around, Benteen cheered his men forward—by madly waving his fishing pole in the air.

When the native force surrendered following the Battle of Bear's Paw soon after, Chief Joseph of the Nez Percé asked if he could meet the striking American officer whom he remembered seeing on the battlefield wearing a buckskin coat, and armed with a pipe and a fishing rod. Benteen introduced himself, and according to witnesses, the two warriors had a private conversation. Perhaps they talked about fish.

Benteen was eventually court-martialed and convicted for drunkenness on duty, but he protested his innocence until the day he died in 1898. His grave is at Arlington National Cemetery.

As historian Richard Lessner put it, "Frederick Benteen remains, so far as is known, the only American soldier ever to go into combat armed with a fly rod."

The combat-fishing incidents of Pickett, Crook, Custer, and Benteen served as fleeting backdrops for the incredibly tragic and brutal experiences of the

American Civil War and the Indian Wars, conflicts that ushered in an era of epic expansion for the United States.

New cities and settlements blossomed across the Great Plains and Western United States, and millions of immigrants poured into the nation to pursue the American Dream. Vast new lands were consolidated and secured, lands that are now home to tens of millions of citizens, as well as to some of the best-loved recreation, hunting, and fishing grounds in the nation. Transcontinental railroads soon linked both American coasts. New technologies speeded up exploration and commerce, taking the American Industrial Revolution to undreamed-of heights.

It was a revolution that was founded and financed, in part, on the high seas, starting early in the century, in the form of commercial fishing, and in the form of a bloody, dangerous quest for a gigantic ocean creature called the whale.

5
America's Floating Industry

They that go down to the sea in ships, that do business in great waters; These see the works of the Lord, and his wonders in the deep.

—Psalms 107:23

There she blows!—there she blows! A hump like a snow-hill! It is Moby Dick!

—Herman Melville, *Moby-Dick*

On November 20, 1820, at a spot on the Pacific equator a thousand miles from land, an 80-ton sperm whale faced off against a 238-ton American whaleship, the *Essex*.

The whale was quiet and still, as if carefully studying its prey.

What happened next began one of the most horrific ordeals in the annals of commercial whaling and fishing, two extremely hazardous, interconnected traditions that helped build modern America.

The *Essex* hailed from the booming American whaling capital of Nantucket, Massachusetts, and was on the fifteenth month of a mission to travel the world to hunt and kill whales.

Many of the ship's nineteen-man crew, plus its twenty-nine-year-old captain, George Pollard Jr., were scattered around the ocean surface in two small whaleboats, being jerked and dragged around by mortally wounded, harpooned whales, in hazardous, unpredictable death journeys that sailors called "Nantucket sleigh rides." One stray coil on such a ride, wrote Herman Melville, meant "a speechlessly quick chaotic bundling of a man into eternity."

On board the *Essex*, first mate Owen Chase, who had stayed aboard to make repairs, spotted the whale in the distance. He watched the creature blast two spouts through its blowhole, then lunge straight toward the ship, bearing down at a speed of three knots, "coming down for us at great celerity," he recalled.

Both ship and whale were about the same length, roughly eighty-five feet. But the sperm whale was one of the largest predators who ever lived, a highly social and communicative creature that had a secret weapon—a mega–battering ram inside his head.

"The immense forehead of sperm whales is possibly the largest, and one of the strangest, anatomical structures in the animal kingdom," explained a team of twenty-first-century researchers. The thick structure, supported by six solid, bony vertebrae, contained two five-hundred-gallon tanks of "spermaceti oil." The purpose of the structure is a mystery, but it may help channel the whale's buoyancy and vocalizations. And it also serves as a head-mounted, energy-absorbing "boxing glove" for whale-to-whale combat over females and turf, and a ramming tool to smash a large object like a whaling ship and escape unscathed.

It is a mystery as to why this whale decided to charge on a collision course for the *Essex*, but with the ocean boiling with the blood of his tormented fellow whales, perhaps members of his own family, the motive might have been simple, righteous fury. To the first mate Chase, it looked like the creature was "fired with revenge" for his brethren.

Usually, whales didn't fight back. "The taking of one of a school, almost always ensures the capture of another," explained Francis Allyn Olmsted, a medical student who travelled aboard a whaler in 1839, "for his comrades do not immediately abandon the victim, but swim around him, and appear to sympathise with him in his sufferings." Some whalers liked to spear a calf first, knowing that the mother would remain nearby, to be killed at leisure, though with California gray whales the strategy backfired, throwing the mother into a murderous rage.

The whale crashed directly into the *Essex* at full speed and punched a huge hole in the hull, recalled First Mate Chase, with "such an appalling and tremendous jar, as nearly threw us all on our faces." The ship, he recalled, "brought up as suddenly and violently as if she had struck a rock, and trembled for a few seconds like a leaf."

After disappearing under the ship, the whale reappeared, churning the water in a frenzy. "I could distinctly see him smite his jaws together, as if distracted with rage and fury," Chase remembered. When the whale vanished beneath the surface, the crew of the *Essex* frantically manned the pumps and tried to repair the ship's gaping wound.

Sperm whale attacks a whaling ship. (National Archives)

Then the whale reappeared nearby, floating motionless, as if it was stunned by the impact. But then it slammed the water with its tail, snapped its jaws, and "started off with great velocity," remembered the first mate, "coming down apparently with twice his ordinary speed, and with tenfold fury and vengeance in his aspect."

"Here he is!" a voice screamed; "he is making for us again!"

This time, the beast had its head half out of the water and was speeding in at a speed of six knots. A sperm whale was capable of speeds up to twenty knots, and was a fearsome sight to behold, the largest carnivore on earth, powered by a four-hundred-pound heart and a brain bigger than any other being who ever lived. The creature hit the *Essex* straight in the bow, a collision that smashed the front of the whaleship to pieces. The ship started sinking swiftly. The whale vanished again, this time permanently.

Water poured into the wreckage so fast that the crew barely had time to toss some bread, water, and tools into a whaleboat and jump in. A horrified Captain Pollard, who had spotted the commotion from a distance, returned in a whaleboat to ask, "My God, Mr. Chase, what is the matter?"

"We have been stove by a whale," replied the first mate. Other whaleboats gathered around and the shocked sailors gazed speechless at the ruins. Some of them, noticed Chase, "had no idea of the extent of their deplorable situation."

Soon there were just twenty men, three little boats, and scattered pieces of wreckage on the surface of the open ocean. There wasn't another ship in sight in this remote chasm of the Pacific. Captain Pollard wanted to head west for the nearest dry land, the Marquesas and Society Islands, but the first mate and crew were convinced (incorrectly) that cannibals lived there. So they made one of the epic miscalculations of maritime history—they embarked on a journey toward the coast of South America, three thousand miles away. In the days that followed, their food spoiled and the boats drifted apart in darkness and rough seas.

The distance to land was much greater, but maybe they'd be spotted by another ship. They were wrong. As they struggled in the infinite sea, their skin roasted in the blazing sun. Their bread got waterlogged at the same time they got dehydrated without fresh water. A killer whale attacked Pollard's boat. They made landfall at barren Henderson Island, which had almost no water or food sources. Three men decided to stay on the island, and the rest of the sailors climbed into the boats again and set off.

In late December, the whaleboats were leaking, supplies were running out, and whales were stalking the boats like vengeful specters. One sailor lost his mind, stood up in the boat, and shouted for water and a dinner napkin. He collapsed into what Chase called "the most horrible and frightful convulsions I have ever witnessed" and died the next morning.

What happened next was a spectacle so gruesome that Chase confessed, "humanity must shudder at the dreadful recital" of the event. The sailors "separated limbs from his body, and cut all the flesh from the bones; after which, we opened the body, took out the heart, and then closed it again—sewed it up as decently as we could, and committed it to the sea." After cooking the man's remains on a piece of stone, they devoured it.

In the days that followed, the boat resembled a slaughterhouse. Three more men died, and their remains were cut into strips, roasted, and swallowed. "We knew not then to whose lot it would fall next, either to die or be shot," shuddered Chase, "and eaten like the poor wretch we had just dispatched." At first, Chase calculated that three men could survive for seven days by eating one human corpse. The gruesome nourishment didn't last long, and only seemed to make the survivors more hungry. They were nearly comatose with starvation, barely able to speak.

After nine weeks adrift, on February 6, 1821, the four survivors on Pollard's boat decided that if they didn't kill and eat someone in the boat, they all would die. They drew lots. The victim designated was the captain's first cousin, teenaged Owen Coffin, whose mother Captain Pollard had told he would watch over and protect.

"My lad, my lad!" cried the captain, "if you don't like your lot, I'll shoot the first man that touches you!"

He offered to take the boy's place, but Coffin insisted. "I like it as well as any other." He lowered his head on the gunwale so a friend could shoot him. "He was soon dispatched," Pollard later said, "and nothing of him left." It was an act of "gastronomic incest," as one scholar later put it.

On February 18, after eighty-nine days at sea, the three survivors in Chase's boat were rescued by an English brig, the *Indian,* near the coastal waters of South America. Nearly a week later and three hundred miles away, the final two survivors in the other boat, Captain Pollard and Charles Ramsdell, were rescued by the Nantucket whaleship the *Dauphin.*

Pollard and Ramsdell appeared crazed as they were hoisted aboard the rescue ship, still sucking the bones of their departed comrades. At an officer's dinner that night aboard the *Dauphin,* a recovering Pollard told the whole story of the *Essex*'s sinking and their three hellish months in the ocean. One officer wrote the whole story down that night, calling it "the most distressing narrative that ever came to my knowledge."

Of the twenty-man *Essex* crew, just eight men survived. Of those who died, seven had been eaten. The three men who remained on desolate Henderson Island stayed alive for almost four months on scant supplies of water, bird eggs, and shellfish, and were rescued by a passing ship. Captain George Pollard took command of another whaleship, was shipwrecked again two years later on a coral reef, then quit the sea forever. He spent the rest of his life as a night watchman in Nantucket. Every year, on the anniversary of the sinking of the *Essex*, he locked himself in his room and refused to eat any food, in honor of his crew. His first mate Owen Chase became a food hoarder, and eventually went insane.

Thirty years after the sinking of the *Essex*, in August 1851, the whaling ship *Ann Alexander* from New Bedford, Massachusetts, faced a similar, highly rare attack in the ocean off Australia. After smashing three whaleboats, a single harpooned male whale crashed into the ship at a speed of 15 knots (17 mph). The enraged whale punched a hole through the hull of the ship below the waterline, sinking the boat in a matter of minutes. Unlike the tormented survivors of the *Essex*, the men of the *Ann Alexander* were rescued just five days later by a passing ship.

When the rescued crew of the *Ann Alexander* returned to the United States a few months later, no one was more amazed than struggling writer Herman Melville, whose newly published novel, *Moby-Dick*, was inspired by the wrecking of the *Essex*. "Ye Gods!" marveled Melville. "What a commentator is this *Ann*

Alexander whale. What he has to say is short & pithy & very much to the point. I wonder if my evil art has raised this monster."

Though whales are scientifically classified as mammals, not fish, the terror and danger of whale hunting came to define the ultimate perils of American commercial fishing. And it helped build the America we live in. As historian Eric Jay Dolin wrote, "From the moment the Pilgrims landed until the early 20th century, whaling was a powerful force in the evolution of the country," he writes. "Much of America's culture, economy and in fact its spirit were literally and figuratively rendered from the bodies of whales."

From the 1600s through the turn of the twentieth century, whale hunting endured as an iconic American industry, at one point representing the fifth-largest industry in the nation. Great family fortunes were made, many thousands of Americans were employed, and the revenues from whaling, along with commercial fishing, provided a critical economic foundation for the economy of New England, and for a time, the great western city of San Francisco.

The Pilgrims were indeed fascinated by the many whales that came to frolic around the *Mayflower*, but they were annoyed that they couldn't catch them. They just didn't have the gear. As one passenger wrote, "Every day we saw whales playing hard by us, of which in that place, if we had instruments and means to take them, we might have made a very rich return."

Soon the Pilgrims adopted the Native American practice of drift whaling, or harvesting dead or stranded whales that washed up on the beach. Lookouts patrolled the coast, and when a whale was harvested, the blubber was boiled in iron cauldrons called "try-pots" and the profits divvied up among the townsfolk. By the 1640s, English settlers in Long Island joined in the trade.

The drift whaling practiced by New England settlers in the 1600s was followed by shore whaling, the hunting of whales near the beach with small boats and harpoons. Bigger ships roamed farther out in the world's oceans—single-masted sloops were followed by two-masted schooners, then square-rigged brigs and steamships. Nantucket Island and New Bedford in Massachusetts became the whaling capitals of the nation, and late in the 1800s, San Francisco came to dominate the trade on the West Coast.

For the ship owners and agents, whaling was a money machine. Whale parts and oil derived from whale blubber were processed and sold to lubricate machinery, and to make candles, house lamp oil, carriage springs, corset stays, luggage and hat frames, buggy whips, umbrella and parasol ribs, hairbrushes, and perfume ingredients. Whale oil made an excellent fuel for streetlamps, which boomed in London in the 1730s as an anti-crime tactic. Whale oil, noted historian Dolin, "lit the world and greased the gears of the industrial revolution." The uses for whale products reflected the growth of America itself, creating fashion and fragrances amid the surging demand by women for leisure products; providing lighting, paints, and varnish for the huge amount of new homes being built; and greasing the factory gears of the American Industrial Revolution.

Whaleships, in fact, took the complex assembly-line factory processes of the Industrial Revolution far out into the ocean: they were floating factory-slaughterhouses that used industrial assembly-line techniques. Historians at the New Bedford Whaling Museum described how the process worked: "Even at the height of New Bedford's whaling prowess in the mid 19th century, the basic procedure remained essentially unchanged: ships were sent to the various whaling grounds with foreknowledge of the seasons when whales could be expected to be present; lookouts were posted aloft; when whales were spotted boats were lowered in pursuit; barbed harpoons were used to fasten to the whale; the harpooned whale dragged the boat through the water until it tired out, whence it was dispatched with a lance. The carcass was towed to the mother ship, where it was cut in (butchered), the blubber tried out (rendered into oil), and the whalebone (baleen) cleaned and stowed; after which the hunt would resume."

The main target of the nineteenth-century Yankee whalers was the sperm whale, the sixty-foot-plus-long, deep-diving, enormous-toothed behemoth that was found in both the Atlantic and Pacific Oceans and offered a potential jackpot of profit for a ship owner. Other whales stalked by American whalers included the Arctic-based bowhead; the right whale, which could weigh up to 100 tons and was easy to hunt, blubber-rich, and once common to the Atlantic; as well as humpback and gray whales.

The New York City–born whaleship sailor-turned-author Herman Melville turned his experiences into *Moby-Dick, or the Whale*, a book of fiction about the hunt for a sperm whale that became arguably the greatest work of American literature. "To produce a mighty book, you must choose a mighty theme," Melville wrote, and in his hands the theme of the whale hunt became a cosmic meditation on struggles familiar to many Americans at the time—the battle of man versus nature, and mankind's search for its destiny in the universe. "All men live enveloped in whale-lines,"

Pursuit of the Sperm Whale, 1856. (New York Public Library Digital Collections)

wrote Melville. "All are born with halters round their necks; but it is only when caught in the swift, sudden turn of death, that mortals realize the silent, subtle, ever-present perils of life." Unfortunately for Melville, the book, first published in 1851, became a bestseller and classic only after he was long dead.

By the 1800s, American square-rigged 300-ton whaleships were roaming increasingly distant waters in search of whales. For the sailors, it was a backbreaking, stinking, bloody business. They hauled the carcass next to the ship with ropes and pulleys. Then they sliced off the blubber with fifteen-foot-long slicing poles into long pieces as the whale was rotated in the water, "precisely as an orange is sometimes stripped by spiralizing it," wrote Melville. The pieces were chopped up and boiled in onboard brick try-pots to extract their oil, which was then stored in barrels belowdecks.

One whaleboat sailor compared the "blood-stained decks, and the huge masses of flesh and blubber lying here and there, and a ferocity in the looks of the men, heightened by the red, fierce glare of the fires," to a scene out of Dante's *Inferno.* According to historians at the New Bedford Whaling Museum, "Processing a whale was nearly as dangerous as hunting one. The deck became so slick with blood and oil that a man could slip overboard to the sharks below. Others were crushed by the enormous weight of strips of blubber or wounded by cutting tools. As the blubber was being rendered in the tryworks, a wave sometimes rocked the ship and splashed scalding oil onto the crew. On rare occasions, the fire in the tryworks spread and devastated the ship. And throughout the days and nights of work, an unforgettable stench clung to the men and their ship." If the trip was successful, a sailor earned a small share of the profits. If not, he was billed for his clothing and food, and often thrown into debt.

Powerful new killing technologies emerged—like the bazooka-like, shoulder-mounted rocket guns that appeared in the 1820s; the Greener gun, a bow-mounted, swiveling harpoon cannon that was introduced in 1837 and was popular through the century; and the revolutionary toggling harpoon, which was invented in 1848 by Lewis Temple, a free African-American, New Bedford–based blacksmith.

The "Golden Age of Whaling" ran from around 1815 to 1860. The United States ruled the waves, with 735 of the 900 whaling ships that roamed the world's oceans in 1846, hunting for sperm, right, bowhead, gray, and humpback whales. At the peak in 1853, American whaleships killed some 8,000 whales worldwide.

Americans dominated the industry because they were the best whalemen in the world. "They sailed the world's oceans and brought back tales filled with bravery, perseverance, endurance, and survival," wrote Eric Jay Dolin. "They mutinied, murdered, rioted, deserted, drank, sang, spun yarns, scrimshawed, and recorded their musings and observations in journals and letters. They survived boredom, backbreaking work, tempestuous seas, floggings, pirates, putrid food, and unimaginable cold. Enemies preyed on them in times of war, and competitors envied them in times of peace. Many whalemen died from violent encounters with whales and from terrible miscalculations about the unforgiving nature of nature itself."

As early as the 1820s, in an era when millions of Southern blacks suffered in the chains of slavery, the New England–based American whaling industry offered a rare avenue of equal treatment for African-Americans, who often earned equal pay for equal work aboard ship. One black sailor reported, "A coloured man is only known and looked upon as a man, and is promoted in rank according to his ability and skill to perform the same duties as a white man." Some black men became officers on whaleships, and a few became captains, like Nantucket-born Absalom Boston, skipper of the whaleship *Industry*, staffed by an all-black crew. He later became a prosperous merchant and landowner on the island. But for any sailor, whaling was often hellish work, with 20-cent-per-day salaries and miserable conditions. The food was wretched, leading one captain to describe it as "beef and bread one day, and bread and beef the next for a change." A voyage could last four years. "We have to work like horses and live like pigs," despaired one sailor in his diary. "It is the most dogish life," one whaler wrote to his brother in 1844. The conditions were "black and slimy with filth, very small, and as hot as an oven." Thomas Roe, a sailor aboard the whaleship *Chelsea* in

1831, described his colleagues as "the most filthy, indecent and distressed set of men I ever came across."

The income from whaling helped build the commercial foundations of cities and towns across New England. The ultimate whaling boomtown was New Bedford, which by 1850 had the highest per-capita income in the United States, and possibly the world. "Nowhere in all America," wrote Melville, "will you find more patrician-like houses, parks, and gardens more opulent, than in New Bedford," "all these brave houses and flowery gardens came from the Atlantic, Pacific, and Indian Oceans. One and all, they were harpooned and dragged up hither from the bottom of the sea." With the building of the transcontinental railroads after 1869, whale oil could be shipped from coast to coast, which launched a thriving Pacific whaling fleet based in San Francisco, with outposts in Alaska appearing in the 1880s.

As American whaleships ranged farther north in the Pacific in search of bowhead whales, they faced the new danger of unpredictable weather combined with ice. In 1871, thirty-two American whaleships were trapped in early Arctic ice off the north Alaskan coast. Incredibly, all 1,219 people on the ships, including some women and children, were safely evacuated by small boats across sixty miles of ice-choked waters to safety. In the 1880s, new steam-powered whaling ships enabled whalers to travel dangerous Arctic waters more safely than sailing ships. The first steam whaler to work in the Alaskan Arctic, the *Mary and Helen*, had an excellent first season—it returned to San Francisco in the autumn of 1880 with 2,350 barrels of whale oil and 45,000 pounds of baleen, a success that the captain credited to the vessel's steam engines.

In 1859, when petroleum oil was discovered in Pennsylvania in 1859, the decline and death of the American whaling industry was just a matter of time. By the turn of the twentieth century, after a temporary rise in demand for baleen whale products between 1875 and 1900, it had gradually shrunk to a small scale to service some women's clothing products like corsets and stays, and when fashions changed, the American whaling industry, which had helped build the nation itself, and symbolized the farthest, most hazardous edge of the young nation's commercial exploration, vanished into the fog of history.

One day in 1924, one of the last American whalers, a ship called the *Wanderer,* set off from New Bedford on its final voyage. It didn't even make it out to sea. In front of a crowd of onlookers, it got stuck in the shallows on a sandbar and wrecked itself. In the twentieth century, other nations like Japan took up the

San Francisco Whaling Fleet, 1864. (Library of Congress)

whale hunt, with increasingly large and sophisticated high-tech factory ships that proved so deadly that several species were hunted to the edge of extinction by the 1960s and 1970s.

But the story is not all tragic. International efforts to protect whales have resulted in some hard-fought victories, some of them close to American shores.

Today, off the California coast, whales also are making a historic comeback. Gray whales, once on the edge of extinction, have enjoyed a big recovery, with a population of more than twenty thousand in 2015.

"Right now, it's a good story—a population that recovered and is doing well," said Wayne Perryman, a federal marine biologist. "The animals look robust and healthy." Whale tour boat operators are delighted. "This was the most impressive gray whale season that I've had in all my years," Captain Rick Powers of Bodega Bay, California, said in 2015. He's been out leading tours for thirty-one years. "We saw gray whales every single trip this season. It's very unusual to go out every trip and bat a thousand." Humpbacks have made a good recovery, too, and they are now routinely spotted off the coast of British Columbia. The once-endangered and overfished populations of harbor seals and sea lions in that region are roaring back, too, thanks in part to the 2009 ban on West Coast krill fishing within the U.S. two-hundred-mile territorial limit, which provides more food for marine life, including endangered fish and birds.

In a happy reversal of the previous order of nature, Americans are now watching whales—instead of killing them.

A perilous life and sad as life can be,
Hath the lone fisher on the lonely sea,
In the wild waters laboring far from home,
For some bleak pittance e'er compelled to roam!

—Anonymous

We have lingered in the chambers of the sea
By sea-girls wreathed with seaweed red and brown
Till human voices wake us . . . and we drown.

—T. S. Eliot

At the same time that whaling blossomed as an American business, commercial fishing for other sea creatures did, too.

The four-hundred-year-plus American tradition of commercial fishing by European settlers in America was first forged in the Massachusetts port city of Gloucester, America's oldest active seaport. The city and its New England neighbors became a global fishing capital because of their closeness to lush, nutrient-rich underwater geography that was the perfect home for bottom-dwelling fish like cod, haddock, and flounder.

Gloucester's cod fishing industry flourished in the 1700s and 1800s, and a full-scale fishing navy of four hundred fishing boats was based at its docks. With success came tragedy, and between 1866 and 1890, more than 380 schooners and 2,450 fishermen never came back from the sea. "The history of the Gloucester fisheries has been written in tears," wrote an anonymous reporter in 1876. In 1882, Captain Joseph Collins of Gloucester asked, "When will the slaughter cease?"

Winslow Homer, The Herring Net, 1885.
(Google Art Project)

By one count, at least 5,000 Gloucester fishermen have died at sea since 1716. The worst year was 1879, when 268 men were lost aboard ships sunk by storms. In just one storm, on the evening of February 24, 1862, fifteen Gloucester vessels went "down to the sea," killing 120 men, leaving widows and 140 fatherless children. Beyond New England, new commercial fishing empires grew and flourished around New York, the Atlantic and Gulf Coasts, California, and Alaska.

On the Chesapeake Bay and Potomac River, commercial fishing gave rise in the 1800s to a powerful breed of American fishermen—the "Black Watermen."

Nantucket fisherman returning home with his catch. (Library of Congress)

They were black men who worked as seamen, boat-builders, sailors, freight haulers, and laborers in the oyster, crab, and fishing businesses. They helped provide the backbone of the Maryland seafood industry ever since 1796, when the federal government started issuing Seamen's Protection Certificates confirming that the holders, including black seamen, were officially American citizens.

The black watermen were among America's first legal black citizens, and their proud tradition of liberty, independence, and largely equal treatment endured through the Civil War and through the worst days of

Gloucester fisherman, early twentieth century.
(Library of Congress)

segregation into the twentieth century. Similar traditions of black fishermen flourished in Louisiana and other spots in the nation. Out on the water, skilled labor of any kind, black or white, was so highly prized that both races were often treated the same.

In 1838, the famed abolitionist Frederick Douglass first escaped from slavery by borrowing the papers and clothing of a free black seaman, and he then headed for Baltimore by train, at one point working as a ships caulker. He noticed that black and white ship carpenters worked together as equals, and that black watermen were highly trusted and respected. When slaves used the Underground Railroad to travel north, Douglass explained, they often preferred water routes, as they "were less likely to be suspected as runaways: we hoped to be regarded as fishermen."

The black watermen were secret heroes of the fight against slavery. "The watermen, mostly black and some white, were the soldiers of the Underground Railroad," explained author James McBride. "Watermen were like cowboys, only more rugged, physically stronger, and tougher and wouldn't hesitate to pull a pistol if they needed to." Before and during the Civil War, the watermen helped escaping slaves, and provided intelligence for fellow African-Americans and for Union forces.

By the 1860s, the Chesapeake Bay was both the biggest source of oysters in the country and a large supplier of shad, creating the need for a strong force of laborers. "Enclaves of free blacks eventually became full-fledged African American communities," wrote historian Harold Anderson. "Along with harvesting oysters, there was work on vegetable farms and in the

Trolling for bluefish. (Library of Congress)

canning, preserving and food packing industries that grew in tandem with oystering and agriculture on the Eastern Shore. For most of the 20th century, blacks comprised the majority of workers in oyster and crab processing houses. Men, women and children worked year round canning tomatoes and vegetables, picking crabs and shucking and packing oysters." With the growth of big cities, demand for oysters boomed, and shucked oysters were canned or ice-packed for shipment to far-flung destinations in America. By 1870, there were more than one hundred packinghouses in Baltimore alone, employing Americans of all backgrounds, including immigrants from Eastern Europe. "Baltimore lay very near the immense protein factory of Chesapeake Bay," wrote H. L. Mencken in *Happy Days* (1940), "and out of the bay it ate divinely."

The tradition forged by the black watermen continued through the twentieth century, when black workers and fishermen worked in the Chesapeake Bay's fish and shellfish industries, and black boat captain-entrepreneurs piloted their own fishing and charter boats. One of them, a man named George Walters, explained: "Many people don't realize it but hand tonging [oysters] was possibly one of the few if not the only industry where there was little or no discrimination. The opportunity was equal to the black man as well as the white man. Some of (Kent) Island's best tongers with the highest incomes were of the black race. Regardless of color, one could work as many days and hours as he wished within the conservation laws and regulations." Another twentieth-century black waterman, Captain Sam Turner, declared, "There ain't no color line out there on that river."

On the other side of the nation, commercial fishing in the Pacific Northwest, especially salmon fishing and processing around the Columbia River Basin, provided employment for many thousands of immigrants in the latter part of the nineteenth century. At first, Chinese men provided much of the muscle work, and they were in high demand for their skill—an expert cutter could clean 1,700 fish per day. When the Chinese Exclusion Act of 1882 prevented new Chinese workers from entering the nation, their place was taken by workers and fishermen from other ethnic groups, including Japanese, Filipinos, Koreans, Scandinavians, Italians, Greeks, and Portuguese. Women were hired, too, often the wives of fishermen. The Columbia River commercial fishing industry peaked in 1883 and 1884, when the catch topped out at more than 42 million pounds, and over 620,000 cases of salmon were packed each year. A thriving commercial fishing business flourished in Alaska, too, based on salmon, halibut, pollock, herring, shrimp, and crab.

By 1880, there were forty-three major fisheries around the United States, providing employment for more than 130,000 people as fishermen and onshore workers. The biggest operations were the Atlantic cod, Chesapeake oyster, and Columbia River salmon fisheries. "Whole communities grew around these industries, often built by immigrants who used their knowledge of fishing to build successful enterprises," wrote historians at the Smithsonian Institution. "Though these fisheries continued to thrive for decades, by the end of the 20th century, each was in crisis."

New technologies have powered commercial fishing ever since the early nineteenth century, when a Cape Ann, Massachusetts, fisherman noticed that mackerel loved to bite shiny hooks, an observation that launched the "mackerel jig," which nearly wiped the species out in the years before the Civil War. In the 1840s, the purse seine, a long net stretched between two small boats to surround a school of fish, was pulling in multitudes of mackerel, herring, and anything else in the way. "The waste during the seining season is enormous, many more being taken off than can possibly be cured, so that hundreds of barrels are left to rot upon the beach," wrote an observer in 1849, who saw "for miles around, the water is completely covered by a thick oily scum, arising from the decaying fish." The twentieth century saw the coming of the industrial-strength, steel-hulled fishing trawler, which towed deep nets that ripped up everything in their path, including fish, sea plants, and the ocean floor, damaging huge areas of marine life.

But even new technologies could not guarantee safety from the savage seas. On the black night of April 2, 2001, the ninety-two-foot fishing trawler *Arctic Rose* vanished from the surface of the desolate Bering Sea, more than two hundred miles from land. The boat was never seen again, and all fifteen men aboard were lost. The body count was so high that it represented the worst U.S. fishing disaster in fifty years, but boats slipping beneath the ocean are tragically routine in the waters off Alaska, where the seas are savage, with constant fierce winds, thick ice, tall waves, and pulverizing cold.

Commander Craig Gilbert, captain of the *Storis*, a 230-foot Coast Guard cutter that protects the Alaska fishing fleet, has sailed both off Alaska and the East Coast. He said the Bering Sea "is by far the worst weather I've ever seen." Of the weather, he reported, "It is almost hard to describe. Brutal. Unpredictable. Remote. Take your pick. People think you're exaggerating. It'll blow for four or five days at 80 knots [92 mph]. Up here they don't call that a hurricane. They call it windy." Winters on the Bering Sea mean 40-foot seas, 100-knot winds, and wind chill of 40 below zero. The water is so cold that it will knock you unconscious within fifteen minutes.

Alaska is now home to the largest commercial fishing port area in North America, Unalaska, which is also called Dutch Harbor. A billion pounds of fish are landed there every year. The top catch is pollock, the most widely consumed fish in the world, which provides the mild taste of everything from McDonald's Filet-O-Fish sandwiches to fake crab sticks. The pollock fisheries off Alaska provide nearly 40 percent of the entire seafood catch of the United States.

The story of American commercial fishing is the story of the uncommon courage and toughness of the fisherman. It is arguably the most dangerous job in America, an often wet, freezing, brutally backbreaking job with hazards including vessel disasters, falling overboard, machine accidents, fog and ice, horrific wind and weather, and rogue waves. The death rate is thirty-one times greater than the national workplace average. Over half of the deaths are caused by a vessel disaster, and another 30 percent involve fishermen falling overboard.

It is also the story of a great American business—the commercial fishing industry today generates over $140 billion in sales and supports 1.3 million jobs in states like California, Massachusetts, Florida, Washington, and Alaska. These and other great American states like New York, Connecticut, Maryland, Louisiana, Texas, and Oregon were built in part on commercial fishing. As historian W. Jeffrey Bolster noted, "Today's fishermen are descendants of the oldest continually operated business enterprise in the New World, one predicated on renewable resources, and one with a centuries-old history of conversations about conservation."

And it is also the story of a long series of boom-and-bust cycles of overfishing, near extinctions, pollution, habitat destruction, recovery, and growth for wildlife in the American waters.

Many years of overfishing in the Southeast and New England have pushed once-healthy species like Atlantic cod, goliath grouper, wild Atlantic salmon, and Atlantic halibut toward commercial extinction. The salmon population in California and the Pacific Northwest has declined to a small fraction of its peak in the 1800s. Tuna fishing off the shores of San Diego, which once boasted a tuna fleet of 160 vessels and was considered the "tuna capital of the Pacific," collapsed in the 1970s in the wake of severe overfishing and environmental concerns.

Decades of abuse by factory trawlers and long lining—in which one boat drags miles of line studded with thousands of hooks—have destroyed sections of the North Atlantic feeding grounds. "We have devastated cod by overwhelming their ecosystem," wrote professor and marine conservation biologist Callum Roberts in 2010. "In our pursuit of fish we have transformed the leafy glades and rolling forests of the sea into endless muddy plains." He added: "The twentieth century heralded an escalation in fishing intensity that is unprecedented in the history of the oceans, and modern fishing technologies leave fish no place to hide."

In 1976, the U.S. Congress launched a major attempt to end overfishing and support American commercial fishermen by passing the Magnuson-Stevens Act, which created an American-only fishing zone of two hundred miles from shore, extended from the previous three-mile limit. But for a while the law had the opposite effect—foreign overfishing was replaced by American overfishing.

Today, thanks to the Magnuson-Stevens Act, public pressure, and more responsible fishing industry practices, the United States is on track to end overfishing for good, and the instances of overfishing and the number of overfished stocks are at all-time lows. In Alaska, America's number one commercial fishing state and a fishery that is in many ways a model of good management, almost no species are on the overfishing and overfished lists. Unsustainable fishing practices have been reduced in many areas of the United States, and many fish stocks have rebounded to healthy levels, including stocks like the Southern Atlantic Coast black sea bass, and the Sacramento River fall Chinook salmon.

Why are men and women so drawn to the dangerous, mysterious ocean? Why do they risk their lives to sail and hunt in deep water?

President John F. Kennedy had one theory, which he explained to an audience gathered for the America's Cup in Newport, Rhode Island, on September 14, 1962: "I think it's because in addition to the fact that the sea changes, and the light changes, and ships change, it's because we all came from the sea. And it is an interesting biological fact that all of us have in our veins the exact same percentage of salt in our blood that exists in the ocean, and, therefore, we have salt in our blood, in our sweat, in our tears. We are tied to the ocean. And when we go back to the sea—whether it is to sail or to watch it—we are going back from whence we came." The famed oceanographer Jacques Yves Cousteau once wrote, "The sea, once it casts its spell, holds one in its net of wonder forever."

According to fisherman William McCloskey, "The work of pulling creatures from the sea for a living is often dangerous, nearly always uncomfortable. Why do men do it? Most would answer honestly: money, like any other work." For some, it may be explained by a powerful sense of wanderlust, as Herman Melville confessed in *Moby-Dick:* "As for me, I am tormented with an everlasting itch for things remote. I love to sail forbidden seas, and land on barbarous coasts."

"As long as there's one fish left," wrote Joseph Garland of Gloucester, Massachusetts, "someone will risk their life trying to catch it."

Every day, across America, the fishermen go out to sea. It is where they test their limits, do battle with nature, and perhaps get closer to God as they do it. They have been doing it since the days of Native American dominion and the early European settlements, through the tremendous growth and expansion of the republic and the dark days of overfishing, and they will probably be doing it as long as there are fish in the rivers and ocean.

They do it for love, for money, for family, and maybe for reasons they can't explain.

They do it because it is who they are.

6
The Golden Age of American Sportfishing

Never mind if the trout aren't biting or the salmon aren't running. They did yesterday. They will tomorrow. If there's one commandment all fly fishermen believe in, it is the optimism of tomorrow.

—Tad Bartimus, Pulitzer Prize–winning writer

One day in 1890, a shy, sickly stockbroker named Theodore Gordon went fishing in New York's Catskill Mountains. What happened next would shape the course of American fishing.

Soon after he cast his fly into the stream, Gordon noticed something peculiar. Until now, most American fly casters, including the thirty-six-year-old Gordon, used artificial wet flies, which sank directly into the water. On this day, however, Gordon watched as a trout pounced eagerly on his wet fly, but the fish did it in the moment it floated on the surface, before it sank.

This got him thinking.

Gordon was a self-taught fisherman who learned to tie his flies by reading British fly-fishing literature, and by studying *The American Angler's Book,* written in 1864 by legendary fishing rod maker and "father of American fly fishing" Thaddeus Norris. Maybe, Gordon figured, dry flies, the kind that were designed to float on the surface and were popular in England, would work in American waters. Gordon wasn't the first to think of this idea, as dry flies were already available on the American market, but he would soon become their champion.

Gordon sat down and wrote a letter to Frederic M. Halford, a famous British dry-fly angler and writer, asking for advice. Halford was a master of making dry-fly designs that matched the look of the hatch of various insects, the kinds that fish loved to wrap their mouths around. Halford was happy to respond by sending Gordon a packet of his custom-made dry flies. Gordon took them into the mountains.

The trouble was, the flies didn't work. The Catskills fish weren't interested in them.

This gave Theodore Gordon another idea.

For European explorers and settlers, the Catskill Mountains were America's first wilderness. Dutch explorer Henry Hudson glimpsed their dazzling, gently rolling peaks in September 1609 as he sailed the *Half Moon* up the river that later bore his name. Dutch and British colonists and post–Revolutionary War Americans delighted in the thickly forested regions' waterfalls, gorges, its 3,500-foot elevations and wild fish and game, and its proximity to the booming port city of New Amsterdam, later New York.

"You know the Catskills, lad," said Natty Bumppo in James Fenimore Cooper's *The Pioneers,* "you must have seen them on your left, as you followed the river up from [New] York, looking as blue as a piece of clear sky, and holding the clouds on their tops."

The mountains were so beautiful that they inspired an entire movement in art. Thomas Cole and other artists went to the Mountain House, the nation's first mountain hotel, and painted lavish scenes of wild gorges and mountains in glowing hues of gold, russet, and violet. Asher B. Durand formed the foundation of what became known as the Hudson River School of landscape painters, and trout fishing was on the minds of these early American artists. Historian Alf Evers explained: "The painters who were attracted to the Catskills by Thomas Cole's success took to traveling the mountains with a paintbrush in one hand and a trout rod in the other."

For fishermen like Theodore Gordon, the Catskills and its two thousand miles of streams teeming with trout were a wonderland. According to author Austin M. Francis, "The typical Catskill trout stream and its surroundings were created on a personal scale that intensifies the feelings of privacy and intimacy with nature so highly prized among anglers. They are the perfect size for fly fishing." The famed American naturalist John Burroughs wrote: "If I were a trout, I should ascend every stream 'till I found the Rondout," a gorgeous mountain creek in the Catskills. "My eyes have never beheld such beauty in a mountain stream. The water was almost as transparent as air—was, indeed, like liquid air. You lay down and drank or dipped the water up in your cup and found it just the degree of refreshing coldness. One is never prepared for the clearness of the water in these streams."

The Catskills were the perfect refuge for the reclusive, never-married Theodore Gordon, of whom only a handful of photographs survive. (One of the photos is of him with an attractive and mysterious woman, but we have no idea who she was.) Gordon learned to fly-fish a fly at the age of fourteen, and he seemed to prefer the company of fish more than people. He had a choppy

Indians Spear Fishing, *Albert Bierstadt, Hudson River School artist. (Google Art Project)*

career in finance, was bedeviled by chest trouble, and escaped to the mountains whenever he could, staying in rickety shacks and farmhouses near legendary trout streams like the Neversink and Beaverkill Rivers and Willowemoc Creek.

As he tinkered in the water with Frederic Halford's British-designed flies, Gordon realized why the fish weren't striking. The flies looked like English bugs, not American ones. Plus, they were built for the straight and gentle currents of English chalk streams, not the fast-churning rougher waters of the Catskills. So he

Theodore Gordon, a Founding Father of American Fly-Fishing. (IGFA)

started improvising. He studied insects carefully and experimented with patterns and materials. He worked up his own new designs that imitated local insects like the delicate mayfly.

Gordon's new dry-fly designs worked like a charm—they mimicked local insects and nymphs so brilliantly that they pulled in trout hand over fist. According to the fishing historian Paul Schullery, Gordon "was secretive about his fly tying methods, almost to the point of paranoia." He added that Gordon "had a nervous energy that was fuelled by hand-twisted cigarettes and he took the odd glass of spirits to bolster up his morale, but his character remains elusive." According to a friend, Gordon was "a cranky old cuss."

When it came to the fragile art of tying flies, Theodore Gordon was a humble artist. "We usually find that [fisher]men of the greatest experience are the most liberal and least dogmatic," he wrote, noting "it is often the man of limited experience who is most confident." He added, "We can never learn all there is in fly fishing, but we can keep an open mind, and not be too sure of anything. It is a fascinating business."

He wrote, "The great charm of fly-fishing is that we are always learning."

Soon Gordon moved to the mountains full time, wrote articles for fishing magazines, served as a fishing guide, and created flies for deep-pocketed city-slicker fishermen for $125 a dozen. Gradually his designs became famous, including the Quill Gordon, which is still hugely popular today. He influenced many great fishing guides, teachers, and fly designers who followed.

By the time Gordon died of tuberculosis in 1915, alone in a mountain shack not far from a trout stream, he had helped inspire and popularize the Catskill school of dry-fly trout fishing designs. He wasn't the first fly fisherman in the United States, far from it. The sport dates at least as far back as October 28, 1764, when the first existing record of fly-fishing was made in America, in the form of a letter from Florida written by a visiting fisherman named Rodney Home. He wrote, "We have plenty of salt water trout & fine fishing with fly in the fresh water rivers of which we have a great number."

But through his writing, his skill, his flies, and his legend, American fisherman Theodore Gordon gave a big boost to dry-fly-fishing in America. In the process, he helped symbolically launch the Golden Age of American Sportfishing, an era that continues to this day.

Fly-fishing is considered by many of its practitioners as the noblest, most elite form of fishing, because of its artfulness, physical complexity, and the serene, understated beauty of the rhythm of casting a nearly weightless imitation fly on a weighted line to a precise spot on the water. It is a paradox: graceful fly movements are created by choppy arm motions. The basics, in fact, are pretty simple. "Flyfishing is traditionally associated with trout, but with the right tackle, fly casters can catch a wide variety of fish species in moving or still water, fresh or salt," explained outdoor writer John Gierach. "But really it's just a matter of casting a fly out on the water (usually at fairly short range), hoping a fish will eat it, and knowing full well that sometimes it won't. That's something anyone who was ever a kid can grasp." The art and science of fly-fishing involves reading the water, finding and catching the fish, and

Fly-fishing in the Adirondacks, nineteenth century.
(Library of Congress)

absorbing the frustrations and spiritual joy of the experience. The other basic methods of sportfishing in both freshwater and salt water are bait fishing, baitcasting, "spinning" with a special rod, reel, and artificial spinner/lure as bait, and trolling, which pulls artificial lures or live bait through the water behind a slowly moving boat.

Just a few years after Theodore Gordon moved up to the Catskills to meet his destiny, an equally historic event happened on the other side of America. It hap-

Trout Fishing Scene, 1870. (Library of Congress)

pened in the waters off California. And it, too, rocked the fishing world.

On June 1, 1898, a distinguished gentleman named Charles Frederick Holder was in a twenty-foot launch off the island of Santa Catalina, fishing with a rod, reel, and six hundred feet of 42-pound test line. His boatman, Jim Gardner, was at the oars. Holder was a naturalist, author, businessman, philanthropist, and former zoologist at the American Museum of Natural History, which his father cofounded. He also was an experienced freshwater angler who fished for striped bass on the Atlantic Coast and fished for the newly discovered tarpon in the Gulf of Mexico. He loved California so much that he moved there.

On this day in the waters of the Pacific, Charles Holder felt a strike, then a pull on his rod, seemingly from a very big fish. Instinctively, the boatman Gardner pushed his oars down into the water to brace the boat. But the leaping fish dragged the two men and their craft off on a wild ride, a battle journey that would change the course of American sportfishing.

The fish was a record-breaking 183-pound bluefin tuna, and before it was finally subdued, it towed the two men for three hours and forty-five minutes, across ten miles of ocean. Back at the Catalina dock, the two men posed for a proud picture with their quarry, which represented the first recorded catch by rod and reel of such a large tuna, the first capture of a large (over 100 pounds) bluefin tuna on sporting tackle.

This was the symbolic moment that modern deep-sea fishing was born. Inspired by his catch, and alarmed by what he thought was the unsportsmanlike practices of other fishermen, Holder and his "Gentlemen Angler" friends established the Tuna Club of Avalon, on Catalina Island. They set up rules for ethical fishing and lobbied for marine conservation, rules that were adopted by fishermen around the world.

The bluefin tuna, once common throughout the At-

Charles Holder and Jim Gardner with the catch that launched the modern age of sportfishing. (Wikipedia Commons)

lantic and Pacific Oceans, is the rock star of the tuna kingdom, and of the deep-sea fish. It resembles a tapered, rocket-powered football that can grow to 10 feet long and 1,000 pounds, with a high-propulsion vertical rear fin that gives it the power to blast through the

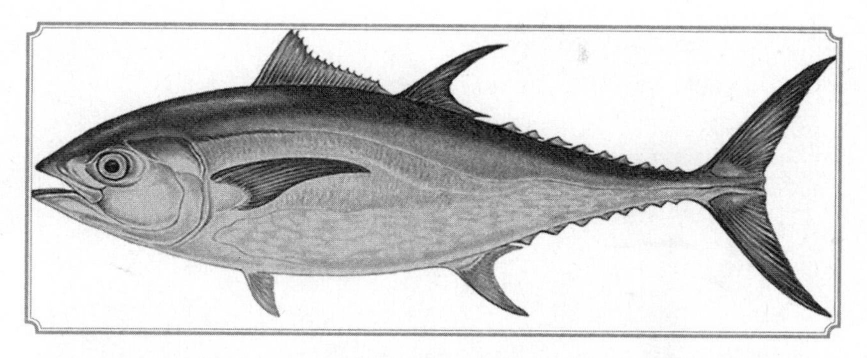

Atlantic Bluefin Tuna. (Massachusetts Executive Office of Energy and Environmental Affairs)

water at more than 40 miles per hour. Like all tuna, they are hard-fighting fish, adding to the challenge for sports fishermen.

"Their sleek muscle-bound bodies cut through water with effortless mastery, driven by a high crescent tail beating side to side in rapid staccato," explained marine conservationist Callum Roberts. "Pectoral fins shaped like hydroplanes flick and twist in the unseen marine breeze, landing remarkable agility to such stiff-bodied creatures."

According to Professor Carl Safina of Stony Brook University, "What allows a tuna to generate such dangerously explosive thrust, merely by wagging its tail, is a package of natural adaptations that exquisitely integrate specialized muscles, specialized circulation, and specialized external design. Making the Bluefin

Tuna Club, Avalon, Santa Catalina, California, the birthplace of American big-game fishing. (New York Public Library Digital Collections)

so unbelieveably tough is a body thoroughly designed to penetrate cold, food-rich waters and rule as the top predator there; its muscles function more effectively in cold water than those of any other fish."

When early-twentieth-century bestselling American author Zane Grey hooked into a then-record-breaking 758 pound bluefin tuna off Nova Scotia, he was amazed by what he saw. "I was struck dumb by the bulk and beauty of that tuna," he wrote. "My eyes were glued to his noble proportions and his transforming colors. He was dying and the hues of a tuna change most and are most beautiful at that time. He was shield-shaped, very full and round, and high and long. His back

A record-breaking catch. (IGFA)

glowed a deep dark purple; his side gleamed like mother-of-pearl in a lustrous light; his belly shone a silver white. The little yellow rudders on his tail moved from side to side, pathetic and reproachful reminders to me of the life and spirit that was passing. If it were possible for a man to fall in love with a fish, that was what happened to me. I hung over him, spellbound and incredulous."

Today, commercial overfishing has radically reduced the populations of both Pacific and Atlantic bluefin tuna (which are in fact, two different species), and the species fell victim, in part, to the booming demand for it as a sushi staple both in Japan and the United States. A number of other tuna species, including albacore, bigeye, and yellowfin, are endangered and often sourced from unsustainable world fisheries.

With the arrival of the motorboat early in the twentieth century, and the invention of a big-game fishing reel with an internal drag by Julius vom Hofe of Brooklyn, New York, in 1913, the stage was set for saltwater sportfishing to take off as an American passion.

By the 1930s, pioneering sportsmen were roaming the seas in custom-built fishing boats, strapped into "fighting seats" with massive mounted rods. The "big game" species are usually considered to include the larger tunas—bluefin, yellowfin, and bigeye; the billfishes: marlins, sailfish, spearfish, and broadbill swordfish; and larger sharks, like the mako, great white, tiger, and hammerhead. Many fishermen add Atlantic tarpon to the list, too. Other popular saltwater fish include bluefish, bonefish, cobia, cod, flounder, grouper, halibut, jewfish, kingfish, mackerel, sea trout, sharks, snapper, and striped bass.

Saltwater fish caught from shore include striped bass, bluefish, tarpon, bonefish, and permit.

There was one saltwater fisherman whose weapons, skill, exuberance, and sheer charisma put him in a class of his own.

He was a big guy who drank hard, fought hard, and spent his life chasing women and fish. He wore a beard.

They called him Papa.

What kind of man goes fishing with a Thompson submachine gun?

Ernest Hemingway, that's who.

If there was one person who captured the epic sweep of the twentieth-century American Sport Fisherman, from old-fashioned cane-pole fishing with a can of worms, to trout fishing with artificial flies and a bamboo rod, to fighting giant marlin from the back

Ernest Hemingway fishing a Michigan creek, age three. (JFK Library)

of a customized party boat, it was the towering literary giant Ernest Hemingway, nicknamed "Papa." Not only was he one of the most revered writers of his era and a global celebrity superstar; Hemingway was one of the founding fathers of sportfishing.

Hemingway fished throughout his entire life, from the streams of Michigan and Yellowstone to the deep salt water of the Gulf Stream and Caribbean Ocean. He hunted for bass, pike, perch, rainbow trout and brook trout, giant black marlin, tuna, tarpon, barracuda, and bull sharks. He hunted whales with a harpoon gun. He used worms, wet and dry flies, and even pistols and shoulder weapons to fish with. He set records, won trophies, and earned a Pulitzer Prize for perhaps the greatest book ever written about a fisherman, his 1952 masterwork, *The Old Man and the Sea.*

He grew up in the Chicago suburb of Oak Park, Illinois, and on his third birthday his father taught him how to fish with a cane pole and a can of worms. On that day, his first time fishing, his mother wrote proudly in his scrapbook that Ernest "caught the biggest fish of the crowd." Forever hooked, he spent countless hours as a boy wandering around the crystalline waters of Michigan, classic trout streams and passages with names like Walloon Lake and Horton's Creek, and the Boardman, Pigeon, and Black Rivers in the wilds of Michigan's Pine Barrens.

In Hemingway's short story "Now I Lay Me," the semi-autobiographical character Nick Adams remembered, "I would think of a trout stream I had fished along when I was a boy and fish its whole length very carefully in my mind, fishing very carefully under all the logs, all the turns of the bank, and the deep holes and the clear shallow stretches, sometimes catching trout and sometimes losing them."

As a teenager, Hemingway took two epic fishing and hiking trips with his buddies in 1915 and 1916.

They packed up pup tents and fishing rods, rode the rails, and hiked the pine forest trails, living on beans and fresh-caught trout for a week at a time. “It is wild as the devil and the most wonderful trout fishing you can imagine,” he wrote. “All clear, no brush and the trout are in schools.” Of the area near Lake Superior, he wrote, “God that is great country.” In his diary, Hemingway joyfully recorded a successful fifteen-minute fight with an 18-inch trout.

On one of these teenaged fishing trips, the seeds of a great writer were born. Hemingway made a list of the people, towns, and rivers he had seen, and jotted in the diary an idea that would eventually shape the course of American literature—he wrote that he had collected some “good stuff for stories.” He wrote a poem about his passion for fishing: “When you started before daybreak, / Mist a’rising from the water, / When your oak strokes sped the row boat past the reeds, / When the line trailed out behind you, / Then a splash! The bass broke water. / He had struck it right beside you. / Tell me brother, / Was not that the day?”

On the day of his first wedding, in Horton Bay, Michigan, in 1921, Hemingway was late for the ceremony because the fishing on the Sturgeon River was so good. That same year, aboard ship in a harbor in Spain, he experienced a revelation that made him want to become a saltwater fisherman: a six-foot tuna "leaped clear of the water and fell again with a noise like horses jumping off a dock." If anyone was good enough to land such a fish, he figured, that person would "enter unabashed into the presence of the very elder gods."

When he was a struggling young writer in Paris in the 1920s, Hemingway's thoughts often wandered back to fishing in Michigan, in the mind of character Nick Adams. "He loved it (the summer) more than anything," Hemingway wrote. "It used to be that he felt sick when the first of August came and he realized that there were only four more weeks before the trout season closed. Now sometimes he had it that way in dreams. He would dream that summer was nearly gone and he hadn't been fishing. It made him feel sick in the dream, as though he had been in jail."

Hemingway had a sad experience with his fishing gear, though, a story told by his own son Jack. "Dad had a big nice trunk," he recalled, "outfitted with all his rods and reels and tackle, everything he had collected over his years of fly fishing. Right after that wonderful summer on the Clark's Fork, in the fall of 1939, he had the trunk shipped by railway express, and they lost it. He never trout fished again. I think it was as if he felt he'd have to start all over again and it just wasn't worth the trouble."

By 1934, having published classics like *The Sun Also Rises* and *A Farewell to Arms,* Papa Hemingway was rich and world famous, and he decided to spurge on a thirty-eight-foot, black-painted, modified Wheeler Playmate cabin cruiser. He named it the *Pilar,* and for much of the next twenty-seven years he roamed the Caribbean in it, hunting for big fish. Based in Key West and then permanently in Cuba, the boat became a center of Hemingway's emotional life, his party boat, exercise machine, and a mental incubator for his writing ideas before he became overwhelmed by depression and committed suicide by shotgun in 1961.

"She is a really sturdy boat," Hemingway wrote of the *Pilar,* "sweet in any kind of sea." Author Paul Hendrickson noted of the *Pilar* that Hemingway "lovingly possessed her, rode her, fished her, through three wives, a Nobel Prize, and all his ruin." Hendrickson explained, "*Pilar* represented this little encapsulated existence where for a long weekend, or just an afternoon, he could get away from the pressures of the writing desk." On the vessel, Hendrickson observed, Hemingway "could be a boor and a bully and an overly competitive jerk, and he could save somebody who was in the water swimming from shark attack on that boat, and he could treat people with uncommon kindness on that boat."

The fishing boat was tricked out with jumbo gas tanks, eight bunks, double rudders, a dual-engine setup of 45- and 75-horsepower engines, a small weapons arsenal, a live fish well, room for 300 gallons of water and 2,400 pounds of ice for chilling daiquiris and beer, a top-of-the-line swivel chair for Hemingway to fight the fish from, a record player, and a cruising range of 500 miles. Eventually he added a flying bridge (top deck) with topside controls of pulleys and a car steering wheel (he once fell drunk off the bridge), a wide roller on the stern to help haul in big fish, and "outriggers big enough to skip a ten-pound bait."

Hemingway often fished seven days a week, from 7 A.M. to 7 P.M. He once fished 54 out of 58 days in a single stretch, and he sometimes blasted away at sharks with a Thompson submachine gun as horrified guests looked on. One time, Hemingway drilled his initials into the top of a shark's head with bursts from the Tommy gun. He dumped the shark carcasses onshore, soaked them in fuel, and burned them in a pit.

On one trip, Hemingway's Cuban first mate Carlos Gutiérrez told him of an old fisherman who grappled for days with an enormous marlin on a handheld fishing line, before it was attacked by sharks and reduced to an 800-pound chunk. Hemingway was captivated by the marlin. It was the Lamborghini of fish, a sleek, powerful, saber-nosed, race-swimming machine capable of 50 mph cruising speeds and weights of 1,800 pounds. Hemingway thought about the story for the next sixteen years and it eventually inspired his Pulitzer Prize–winning masterwork, *The Old Man and the Sea.* Like Herman Melville's *Moby-Dick,* one of its chief competitors for the title of "greatest American novel of all time," *The Old Man and the Sea* explored the theme of man's struggles in nature, and with himself, through the story of an obsessive quest to conquer an ocean creature.

One day, Hemingway took a Jesuit priest friend named Father J.S. McGrath out on the *Pilar* for a fishing trip in the Gulf Stream. His belly full of frosty beer and sandwiches, Hemingway was in a great mood. As he expertly piloted the boat, the Great Man announced his fisherman's drinking philosophy: "Sun and sea air, as they dry your body, make for almost effortless beer consumption. The body needs liquid of a nourishing kind. The palate craves coolness. The optic nerve delights in the sensation of chill that comes from its nearness to the palate as you swallow. Then the skin suddenly blossoms with thousands of happy beads of perspiration as you quaff!"

Suddenly, the priest cried out. A billfish had struck his bait.

Hemingway and friend with marlin on Bimini dock.
(National Marine Fisheries Service)

"Reel as fast as you can, Father!" called Hemingway from the bridge. But a shark swiped the fish from the holy man's line.

Then the priest hooked a gigantic sailfish, which began to jump out of the water no fewer than twenty-eight times, as Hemingway skippered the boat and the priest struggled with the rod.

"Fight him, Father!" hollered Papa.

But after fifteen minutes, the priest's arthritic left hand locked up. "Ernest, you must help me. I can't handle this fish any longer."

"Look, he's yours," said Hemingway. "He's a sail-fish, not a marlin as I first thought. He may be of record size. If I take over, the fish will be disqualified for any kind of record."

"But I can't go on," pleaded Father McGrath.

Hemingway agreed to take over for the priest, fought the fish for fifteen minutes, and finally hauled in a nine-foot, three-quarter-inch-long, 119.5-pound behemoth, a record-breaking catch and the biggest Atlantic sailfish ever landed. Following the strict letter of the rules, Hemingway declined recognition for the catch, and wished he knew of a way to give Father McGrath the credit.

In 1935, Hemingway hoisted the first intact, unmutilated giant bluefin tuna, a 310-pounder, up unto the docks at Bimini, in the Bahamas, the first fisherman ever to do so. Until then, the sharks always got a piece of the tuna first. But Hemingway had perfected an aggressive technique of applying constant "pump and reel" pressure to boat the fish as fast as possible, rather than running it out until it got exhausted. Ever since, the technique has been known as "Hemingwaying" a fish. "The secret is for the angler never to rest," he once explained. "Any time he rests the fish is resting."

Hemingway was a brilliant student and teacher of deep-sea fishing. He invited scientists along on fishing trips to study currents and species, he loved to show newcomers the ins and outs of fishing on the open ocean, and he filled his ship logs with reams of meticulous weather and fishing data.

He won every fishing contest in Key West, Havana, and Bimini and was named lifetime Vice President of the International Game Fish Association, a group that he helped establish. "A big-game fisherman might have counted himself blessed to have landed two or three good-size marlin in a season's fishing," wrote Paul Hendrickson. "In one month alone, May 1932, right after he'd begun, Hemingway had landed nineteen marlin on a rod and reel. The following year, fishing again on the north coast of Cuba, from mid-April to mid-July, Hemingway brought in fifty-two marlin. The largest of these, a black marlin, went 468 pounds and nearly thirteen feet, a Cuban record."

In the cool breezes off Bimini in the summer 1935, Hemingway loaded his Thompson submachine gun and went on a shark-shooting rampage. "Shot 27 in two weeks," he wrote proudly to a friend. "As soon as they put their heads out we give them a burst." He and a friend fought a 1,000-pound tuna for nearly ten hours, "Then just when we had him whipped and on the surface and showing terribly big in the searchlight at 9 o'clock at night 17 miles from where he was hooked, the sharks hit him. 5 hit him at once. I shot 3 with the Mannlicher [Italian-made bolt-action rifle] but they cut him like a log in a planing mill."

One day, Hemingway tried to wrestle a big shark onto the *Pilar.* The shark fought back. In the midst of the struggle, Hemingway whipped out his .22-caliber Colt Woodsman automatic pistol to settle the issue, but he managed to shoot himself instead, with a ricochet off a brass fixture that he reported went "through both legs with one hand while gaffing a shark with the other." He patched himself up with iodine and bandages and escaped to Key West for treatment.

One day in 1938, Hemingway bagged a world-record seven marlin in a single day. He hauled in one of the biggest marlins in history, a giant weighing nearly 533 kilograms. Marlin fishing, he declared, was "truly the most wonderful damned thing I have ever been on."

Ernest Hemingway seemed to be on a lifelong quest to prove he could out-punch, out-elbow-wrestle, out-drink, out-shoot, out-write, and out-fish any other man on earth. In the last two departments, he achieved almost total victory.

Hemingway helped establish the group that became today's International Game Fish Association, or IGFA. In 1950, he organized a competition that continues today as the Hemingway International Billfishing Tournament, a four-day tournament where contestants go for marlin, tuna, wahoo, and other fish using 50-pound fishing line, which was won by Fidel Castro (fair and square) in 1959. A famous photo shows Hemingway presenting the soon-to-be Communist dictator with the silver winner's cup, both men wearing what Hemingway's niece Hilary called "the two most famous beards of their time." In 2015, Americans returned to participate in the Cuban-based contest for the first time in decades.

I'm partial to Hemingway because he was a big muscular guy with a beard who loved to hunt and fish, just like me. He's definitely the kind of guy I'd love to go fishing with. I think we'd get along great.

World War II Emergency Fishing Kit. (Library of Congress)

I might bring along a bulletproof vest, though, just in case Papa decided to open fire with his Tommy gun.

By 1941, American sportfishing was booming. Books and magazines were catering to fishermen, more and more specialized fishing tackle was being sold, and boating was becoming accessible to the middle class. But suddenly, when America entered World War II at the end of that year, the sport went on hold.

Then something interesting happened. The American fisherman pitched in to help win the war, as did the whole country. Fishermen signed up to serve in Europe and the Pacific. Tackle manufacturers switched over to making military supplies. And the fishing industry got together to help servicemen survive at sea.

Fishing experts realized that if you were stuck on the ocean in a raft without food or water, you could survive, as long as you had the right tackle. Not only could you pull up fish for food, but you could extract a drinkable liquid from the fish that could substitute for water. So in 1942, U.S. military planners huddled up with outdoor writers, fishing groups, and tackle makers, and together they came up with something brilliant—the "Emergency Fishing Kit."

The kit was a compact, eight-piece pouch that contained everything you needed to hook fish in a pinch: hooks, line of various sizes, an assortment of artificial baits, dehydrated pork rind, a knife, a hand net, a whetstone, swivels, sinkers, a fish spear, and an instruction sheet on waterproof paper, all sealed up in a metal cylinder. By early 1943, the kits were standard issue, packed aboard thousands of lifeboats and survival rafts used by the U.S. Navy, Army, and Coast Guard.

The kits were highly popular, saved hundreds of downed sailors and airmen during the war, and helped feed thousands more Allied service members in far-flung, remote locations around the world.

"We have had considerable luck, including a king-fish that weighed eighty pounds," wrote one officer. "The kits are of value for recreation and they improve morale." Another wrote, "The men have become Isaac Waltons of the first order." From the South Pacific came the report, "Here on Fiji your kits have brought an assorted variety of fish, some of them weighing thirty pounds." A corporal wrote, "The kits have been a god-send; fish is the chief diet now—and the boys have hooked some whoppers." In the summer of 1944, servicemen stationed on Midway snagged more than 40,000 pounds of fish, including tuna and wahoo, which worked out to two meals a week each for more than 6,000 men.

When the war was over and the boys came home, what do you think many of them did with their new-found angling skills? They became American fisher-men, and they kicked off a huge boom in the sport, one that continues to this day.

The story of the American Fisherman is the story of families.

It's the story of dads and moms, sons and daughters and grandparents going out into nature together and experiencing what it feels to be alive on God's earth, enjoying his gifts.

I know of one family up in Nebraska that's especially fond of fishing. Not only do they love to go fishing; they also sell tens of millions of dollars of fishing gear every year through their retail stores and mail order business. They are the Cabela family, and they are one of America's First Families of outdoor sports. Dick and Mary Cabela, along with Dick's brother Jim, started the business on their kitchen table in 1961. Today they do more than $3 billion in sales every year.

The late Dick Cabela saw fishing as a way of bonding family members together in a shared passion. He once wrote, "It is always great to see your loved ones do well in any endeavor, to see them truly enjoy your passion for fishing, to witness a transformation where your passion becomes theirs and you share something so deeply words are unnecessary. This is when you realize your true purpose is to make someone else's life better."

His wife, Mary, an accomplished hunter and angler, sees fishing as a way of learning life lessons, and of getting closer to the Creator. "Most people begin fishing because someone took them—a parent, a grandparent, or a friend," she told us. "When we first started taking our kids fishing, I think we did it because someone took us out fishing as children and we longed to give that same gift to our loved ones. And fishing is a gift. It is a gift from God and a gift we can pass on that gives far more than the hours of memories it provides. Fishing is a lesson in patience and perseverance disguised as good, clean fun. When you spend time fishing together, it teaches you so many life lessons, but most of all it teaches you how to be a better family member and a better friend. In fishing, like in life, there are many periods of seeming monotony followed by short periods of drama, excitement, fear, failure, and triumph. In fishing, like in life, you must prepare for those moments, and embrace them when they come. Fishing fuels the fire of hope. Just one more cast and that hog might strike. Just one more cast might be the perfect one. One more cast might create the memory of a lifetime. Just one more. I wonder if that is the way God fishes for us. He puts the Word, and the miracles of life, and the love He has right there, hoping for us to take it. Casting time after time after time. Maybe someday we'll be hungry enough to allow ourselves to be caught."

And David Cabela, Dick and Mary's son, thinks that fishing brings us closer to family, nature, and God, all together at the same time. "The face of an angler who has just hooked into a fish, reveals the unfiltered joy of a child," he told us. "Fishing is one of life's great metaphors. You set out full of hope and anticipation and wonder. You focus on your goal, sometimes oblivious to the lessons and beauty in the smallest of moments. You question your tactics. You question the fish. You question yourself, and at times, you even question God. Just when you think you've hooked into the big one that you've been hoping for, your line goes slack. You fail. You succeed. You experience moments of disappointment and joy—sometimes inexplicably at the same time. You work hard for just a chance at a moment you may never forget. Then, at the end of the day, when the sun and the water come together in a soft glow and the peace of a moment truly lived fills your heart, you realize it was never about the size or quantity of the fish. It was not really about catching fish at all. At the end of the day, it is about relationships. Relationships with your friends, your family, with the water, the fish, with yourself, and with God. At the end of the day, when the sun and the water embrace as if they were always one, it is about learning to give yourself to the moment and to accept with gratitude the gifts you have been given."

As someone who has been fishing since he was three years old, I couldn't agree more.

Today, the bass is the undisputed King of American Sport Fish.

The largemouth and smallmouth bass are the most popular, most sought after, and most celebrated freshwater game fish in modern America. Books are written about the bass. Songs are sung about them. Bass fishing has long been popular in the United States, but really took off in popularity after World War II, when millions of servicemen returned home to discover a great leisure sport, close to their own backyard. Today, multimillion-dollar tournaments like the Bassmaster Tournament Trail, organized by the Bass Anglers Sportsman Society (B.A.S.S.), are devoted to the fish. In 2013, the B.A.S.S. alone had 500,000 members. Fishermen obsess about bass, sometimes even to the point of risking their jobs and life savings in the process.

Why does the bass cast such a powerful, addictive spell on the American fisherman? Three good reasons. It leaps high, fights hard, and is everywhere and nowhere at the same time—originally common in the eastern states, it now lives in every state except Alaska, yet can be notoriously elusive. Their excellent hearing, vision, and vibration sensitivity makes them hard to catch, since they may sense you first and flee. They're hardy, muscle-bound, supertough predators that can weigh over 20 pounds and will gobble up nearly anything in their swim path, including crayfish, minnows, worms, leeches, bugs, lizards, frogs, other fish, even

The largemouth bass, Micropterus salmoides. *(New York Public Library Digital Collections)*

water snakes, ducklings, flip tops, bottle caps, mice, and cigarette filters. Like trout, bass love to hang around underwater cover like weeds, logs, boulders, and drop-offs. "You may find structure which at the moment is not holding bass," noted bass expert Buck Perry, "but you will *never* find bass without structure." Striped bass, often common to America's coastal areas, are also a prized marine sport fish.

There are lots of fun ways to fish for bass with bait or with artificial lures like Texas-rigged plastic worms, and lots of places to find them—ponds, lakes, creeks, reservoirs, swamps, and big rivers. Tournament fishermen, who usually practice careful "catch and release" bass, tend to either be "power fishers," covering a lot of water quickly in search of their prey, or slow and patient "finesse fishers." For the regular fisherman, jerk-bait fishing is a simple, easy way to fish for bass, by yanking and jerking your rod tip around so your bait impersonates a wounded baitfish. Bass love to strike crippled targets, so they'll pounce on yours. Similarly, with crank-bait fishing, you tease the bass by fast-reeling, or cranking, your sunken bait, stopping to let it rise, then repeating the move, bouncing it around structures or "bumping a stump" to trigger a reflexive strike by the bass. This technique works well around drop-offs, solid structures, and heavy underwater cover. For top-water bait, there are similar tactics like "popping" a lure across the water to trigger a surface strike.

More bass-fishing methods include dragging a Carolina rig, or pitching, skipping and flipping lures to tricky spots, like under a boat dock. Champion bass angler Mike Iaconelli described his technique for skipping a lure: "When you cast, stop halfway instead of following through, similar to a check swing in baseball. This makes the lure hit the surface of the water a few feet before your target, so the lure skitters over the water. It's a good way to get under docks and other structures."

The story of American sportfishing is a tale of millions of Fish Who Got Away. Every fisherman has one, maybe a hundred such stories. It happens to all of us, from humble beginners to tournament champions.

Back in 1986, a fellow named Jim Bitter from Fruitland Park, Florida, had one of the all-time whoppers of these stories. In one instant, he was on the edge of being the most famous bass catcher in the United States. He had a million-dollar bass in his hands on his boat on the muddy James River on final day of the Bassmaster classic tournament in Richmond, Virginia. The 13-inch fish itself wasn't especially big, but it was big enough that when added to Bitter's tournament catch so far, it would put him over the top. Everyone figured Bitter was about to be the winner. Championship speaking fees, endorsement deals, and media appearances were about to land on his shoulders like angels of fortune.

Then the fish flopped out of the boat. He never saw it again. The clock ran out. Jim Bitter did not win the tournament. "I quickly grabbed my measuring board and placed it on top of the tackle box, which was just about level with the gunwale (side of the boat)," Bitter explained. "When I went to put the fish into the livewell it got out of my grip somehow. The fish bounced on top of the tackle box and did a perfect squirt into the water, like someone had a string and pulled him overboard." He was baffled as to how it happened. "I don't know why I measured him on the tackle box instead of doing it on the floor of the boat. I guess I just got excited to catch the fish that I thought I needed to win."

It was a heck of a long ride all the way back to Florida for Jim Bitter and his wife. "The vision of that one bass bouncing off the tackle box kept playing in my mind all the way home," Bitter said. "I just kept thinking how sweet it could have been."

Bass-fishing megastar and TV fisherman Roland Martin told the story of how he once held a giant 15-pounds-plus largemouth in his hands on Santee Cooper Lakes in South Carolina. It might have been a record-breaker. The creature wriggled free once, Martin recaptured him, then it slipped out again, this time for good.

Jack Chancellor, another B.A.S.S. Classic winner, remembered a 5-pound bass he lost when he was on track to winning the world championship. "When that fish swam away," he lamented, "it looked like it had dollar signs all over it." Paul Elias, winner of the

1982 Classic, had similar memories of a bass that got away during the 1983 Super B.A.S.S. Tournament on Lake Lanier, Georgia. "I was up in the top ten, and had a good shot at winning," he recalled wistfully. "It was on the last day. I had about a five-pound fish on a crankbait. I could see it in that clear water. It was about two-thirds of the way back to the boat, about two feet under the water. The lure just slid out of its mouth for no reason. The fish just turned and slowly swam off. All I could think about was that was about twenty thousand dollars swimming away."

Larry Nixon, a top professional bass fisherman, has never forgotten a giant fish he hooked on his favorite local lake, Toledo Bend in Arkansas. He remembered, "The fish made a move as if coming to the top, but then went under the boat. There was so much pressure on my line, I couldn't mash down the release button on my reel. The fish took out twenty-five yards of line against the drag and broke off on twenty-pound test line. I'll never know how big that fish was. To this day that fish sticks out in my mind."

Just as bittersweet a memory as "The Fish Who Got Away" is "The Fish You Never Caught." Jack Dennis Jr., of Jackson, Wyoming, once explained, "I've guided on the New Fork River for twenty-five years but I've never landed a trophy on her. There's a certain mystery to that river, like wondering whatever happened to the girl you loved in high school but didn't marry."

America has spawned a host of highly skilled and brilliant sportfishermen. They are fathers and mothers, sons and daughters, boat captains, backwoods anglers, and champion deep-sea fishermen.

For nearly a half century, outdoor writer and photographer Bernard "Lefty" Kreh of Maryland has been one of America's greatest fly-fishing teachers and enthusiasts. Roland Martin, Rick Clunn, and Mike Iaconelli are considered among the greatest bass fishermen who ever cast a fly. Other American fishing greats include sportfishermen like Pete Maina, who started as a guide in Wisconsin at age twelve and pioneered catch-and-release for muskies; former Olympic skier Andy Mill, who became a champion tarpon fisherman; steelhead champion Buzz Ramsey; former dentist and top professional walleye angler Gary Parsons; and retired schoolteacher David Pickering of Rhode Island, who catches an amazing average of two thousand striped bass per year from the shoreline.

Another legendary figure in American sportfishing today is Dr. Martin Arostegui, of Coral Gables, Florida, a retired emergency room physician who has achieved more than four hundred world records over the past two decades, in salt water and freshwater. He ties his own flies and has said he safely releases more than 90 percent of his record fish. He has said that being a retired doctor gives him two advantages for catching record fish: experience with precision handwork, and earning enough money to let him do all that fishing!

Tommy Gifford has been called one of the greatest charter skippers and bluewater anglers who ever lived. Operating from Miami, he popularized game fish places like Havana, Wedgeport (in Nova Scotia), Bimini, St. Thomas, and Montauk, was a guide and consultant to great anglers like Ernest Hemingway, and innovated saltwater gear like the flying gaff and the star-drag reel. "Big game angling has a brief history," noted outdoor writer Raymond Camp, "but Tommy Gifford's name is sharply etched on every page." Up on Lake Michigan, another legendary charter captain, Denny Grinold, has been bagging salmon and trout with clients ever since 1973.

Kevin VanDam from Kalamazoo, Michigan, is a towering figure in sportfishing, having become by 2010 the all-time money winner on the Bassmaster Tournament Trail, and winning more than $9 million on the pro angling circuit. Many folks think he's the best bass fisherman who ever lived. ESPN commentator Mark

Champion bass angler Kevin VanDam. (Seigo Saito, B.A.S.S. Photo)

Zona has called him "as close to perfection as I've ever seen." VanDam himself once stressed the sheer complexity of the sport of competition fishing: "There's no question in my mind that fishing is a science. Where the fish are, what they're up to, everything they do is based on environmental conditions. So there are a jillion ever-changing variables. It would be like playing golf, except that the pin moves all the time."

What separates Kevin VanDam from the rest of the pack? "It's the decision making," he once explained. "The mental part of the game. Lebron James when he has a 50-point game or Tiger when he shoots a 63. Or me on the final day of the Classic when I caught 27 pounds—you can feel it, you know it. I wish I could get in that zone or situation every day. All I know is that when it's happening, you know it's going on."

One of VanDam's secrets to fishing greatness is—cookies. His wife, Sherry, bakes him a sack of lucky high-calorie peanut butter Toll House cookies to keep him going through the 5,500-calorie-per-day burn rate of a grueling eight- or nine-hour tournament day. VanDam is definitely an ultraenergetic, aggressive competitor. He has said, "My greatest strength can also be a weakness at times, it's just my style of fishing. I fish fast and cover a lot of water. It does help me locate fish really fast, but sometimes I can go too fast and be too aggressive and not realize how good a spot might have been." One of his tips: "The one big thing that I think would help a lot of people catch more fish is learning the importance of making a quiet cast with a soft presentation. Fish don't have to be hungry to be caught, and being discreet with your casting is a way to get more of these fish."

Today, recreational fishing is a huge, booming business, generating more than $100 billion in economic impact and more than 300,000 jobs, with the top five states for jobs being Florida, North Carolina, Louisiana, Texas, and New Jersey. There are over 46 million anglers in America, which is more than the number of golfers and tennis players combined. There are over 10 million American saltwater anglers alone, and more than 90 percent of them travel from the Atlantic and Gulf Coasts. They take 68 million fishing trips per year. Through licenses, fees, taxes, and donations, American fishermen contribute more than $1 billion a year to protect the environment.

Fishermen are a powerful voice for protecting American waters and the life they contains. "A lot of people talk about conservation," declared *Field & Stream* editor Ken Schultz. "Anglers, responsible anglers, live it. They have an interest in clean, uncontaminated water and avoiding pollutants, and they are intensely aware when nature is out of whack. They pay for conservation, too. Most fisheries resource facilities are financed from fishing license fees." He added that the widely popular practice of catch-and-release fishing is proof of the conservation ethic of the fisherman. "The enjoyment is in the challenge and the accomplishment of catching the fish," wrote Schultz. "Then, let it go. You don't have to fill up your freezer or give a lot of fish to friends to prove you had a good day. More than half, 58 percent, of American anglers now catch and release, according to the latest statistics. Younger people, especially, realize our resources are finite."

If I had all the time in the world and nothing to do, I'd grab some tackle, pack up the family in the RV, and go on a fishing trip. For about three years.

It would be a dream journey, the greatest American sportfishing trip of all time. First, we'd mosey around Louisiana for a spell, fishing in magic fishing spots near home like Henderson Lake, the Caney Creek Reservoir, and the 27,000-acre Caddo Lake on the Texas state line near Shreveport, home to a mind-boggling number and variety of fish. Some folks think it's the prettiest lake in America.

Then we'd branch way out, and travel to fishing nirvanas like Lake Okeechobee and Rodman Reservoir in Florida, Big Spring Creek in Pennsylvania, New York's Catskill Mountains, Lake Fork in Texas, Briery Creek Lake in Virginia, and Georgia's Montgomery Lake, where a twenty-year-old farmer named George Perry pulled out a 22-pound, 4-ounce largemouth bass on a Creek Chub Fintail Shiner lure on June 2, 1932, a record that still stands.

"All at once the water splashed everywhere," fisherman Perry recalled decades later. "I do remember striking, then raring back and trying to reel. But nothing budged. I thought for sure I had lost the fish—that he'd dived and hung me up. I had no idea how big the fish was, but that didn't matter. What had me worried was losing the lure." As he finally hoisted up the enormous 32.5-inch bass in both his hands, Perry thought, "how nice a chunk of meat to take home." The catch provided twelve meals for his family, and $75 worth of prizes from the record keepers at *Field & Stream* magazine—a Browning shotgun, shells, and some clothes. It remains the most-sought-after sportfishing record in history. If someone caught a bass bigger than Perry's today it would be worth at least $1 million in endorsements and promotional value. Better yet, if you kept it alive and healthy and put it on display, you'd make a few million more.

On my fantasy fishing trip, I'd definitely spend some time on that Georgia lake, to see how big the largemouth are these days. Next, my family and I would spend a few months in the RV wandering around the fishing paradises of Wyoming, Montana, and Idaho and see if we could retrace the steps of Lewis and Clark's epic salmon-powered journey through the Rockies.

Finally, I'd conquer my discomfort of the ocean and take the family on a saltwater fishing marathon, through Cape Cod, Long Island, the Jersey Shore, the Chesapeake, the Keys, the Gulf and Pacific coasts.

We'd reel in 8-pound bonefish, 36-inch striper, big striped bass and bluefish, redfish and sea trout, tarpon, bonefish, yellowtail, small tuna, maybe some 50-pound sailfish and 500-pound marlin.

That would truly be one amazing fishing trip.

As soon as the adventure was over, I'd start dreaming about where we should go fishing next.

More than half the intense enjoyment of fly-fishing is derived from the beautiful surroundings, the satisfaction felt from being in the open air, the new lease of life secured thereby, and the many, many pleasant recollections of all one has seen, heard and done.

—Charles F. Orvis

They say you forget your troubles on a trout stream, but that's not quite it. What happens is that you begin to see where your troubles fit into the grand scheme of things, and suddenly they're just not such a big deal anymore.

—John Gierach

The angler forgets most of the fish he catches, but he does not forget the streams and lakes in which they are caught.

—Charles K. Fox

7
Rise of the Great Women Anglers

I followed mountain streams in search of those wary, fighty, cocky little brook trout. There was sport! To follow a brook, to see a "hole" hemmed in by birches or alders, to stand on rolling stones, and cast your line, gently, as you hide in shadow; to feel your fly nipped by a passing flash; to give a quick jerk of the rod, only to find your hook fast in an old log, under which the trout had hidden. That is all pleasure and excitement.

—CARRIE FOOTE WEEKS, *AMERICAN ANGLER*, 1900

I love it passionately. I don't care for clothes; I don't care for jewelry; you may take away anything I have, but I won't give up my fishing. . . .

There is nothing I like so much in all the world, and it is my desire, my ambition, and my wish sometimes to live in a little hut, to be a woman hermit and do nothing but fish.

—A Woman Angler from New York City, quoted in the *New York Times*, August 14, 1898

One day in the spring of 1896, thousands of spectators at New York's Madison Square Garden were witness to an astonishing sight.

There at the State of Maine exhibit at the second annual Sportsman's Show, surrounded by a mob of people, stood a striking, six-foot-tall woman on an elevated platform, whipping a bamboo fly rod toward a giant tank of fish.

"Simply shocking!" gasped an onlooker, gaping at the woman's pale green deerskin skirt, which rose a daring, dangerously risqué seven inches off the floor. Patterned after the latest Paris fashions, the woman's hunting outfit was custom-tailored by the Spalding Brothers' sporting goods company to showcase its new ladies' activewear line.

The hunting costume, raved a reporter from the *Washington Times,* was "probably the most expensive and elaborate ever made in this country, the material alone cost over $100 and cannot be duplicated in this country." The woman with the fishing rod looked like a high-fashion goddess of the wilderness. One dazzled reporter described her as "supple and lithe as a young tree." Around the band of her hat, she hooked her own handcrafted artificial flies, pioneering a sports fashion that endures to this day.

As the huge crowd of men, women, and children gazed up spellbound at the woman, wrote one historian, she "awed the wide-eyed New Yorkers with her piscatorial prowess by repeatedly casting her rod over the tanks and getting strike after strike with a delicate turn of the wrist."

The woman's name was Cornelia Thurza "Fly Rod" Crosby. She was the first Female American Fishing Superstar.

"Nobody in New York had seen anything just like me before," Crosby later quipped.

"Tall, attractive, and modest, with a resonant speaking voice and attired as she was, the effect was electric," reported the *New York Times.* "Crowds flocked, lingering for her lectures and demonstrations. Cornelia's garb served to create a new fashion trend for American women."

If you measure the greatness of a woman by the impact she has on a sport, Cornelia Crosby is one of the all-time giants of American fishing. She was a gifted angler, a crack-shot hunter, a wilderness advocate, a skilled writer, and a marketing genius, who, more than anyone else, turned the great state of Maine into an outdoor sporting powerhouse at the turn of the twentieth century. And as if that weren't enough, as a 1903 article declared, "She was the first champion of woman's rights in the hunting and fishing line, and was the first to advocate that women should go into the woods with rifle and shotgun to enjoy the sport which men had preempted."

Born in the Rangeley Lakes region of Maine in 1854, her father died when she was young. She lost her brother to tuberculosis, and she was weakened by the disease herself for much of her life. In her teens, Crosby worked odd jobs at summer camps and picked up basic fishing and hunting skills from local Native Americans and guides. She was a good writer, and earned a diploma from St. Catherine's Episcopal School for Girls in Augusta, Maine. In her twenties, she worked as a bank clerk and telegraph operator in Franklin County.

As a young woman, Crosby received some devastating news from her doctor: she only had a few years to live. Her lungs were debilitated by tuberculosis, her muscles were weak, and she was anemic. She developed a lung infection. Crosby's only chance to survive, said the doctor, was "large doses of the great outdoors."

In a final attempt to save her life, one June day, some friends carried Crosby to the foot of western Maine's spectacular Mount Blue. "Here at a farmhouse I was to try the healing power of nature," she recalled. "A brook full of trout came laughing down the mountainside, and from there I took my first trout, with an alder pole." She was hooked on fishing for life, and soon she was fishing, hiking, and hunting the wilderness of Maine's Rangeley Mountains with a passion, using the Rangeley Lake House as her headquarters. A friend gave her a prototype lightweight bamboo fly rod, which she loved, and before long she was amazing the local fishermen with her fly-casting skills. One historian wrote that "in Cornelia's hands the fly rod became a magic wand that launched her amazing career."

Crosby, who never married, was blessed with star-quality charisma and a witty, outgoing personality. She was soon locally famous for her outdoor talents, and by the time she reached her thirties, she was on a path to become America's first superstar outdoorswoman.

In the woods, Crosby was said to have a sixth sense for choosing the perfect fly, and for finding the choicest salmon and trout. She befriended the native people of Maine, one of whom said, "Her face is white, but her heart is the heart of a brave." She was said to have once caught two hundred trout in a single day at Kennebago

Lake, and she was reportedly able to knock down distant caribou with precision marksmanship. "She is as patient in whipping a stream in the Maine woods as any man," raved one journalist, "and when she gets a strike, is far more expert then most men who pride themselves on the dainty way in which they can kill their game." Another journalist wrote, "She is an expert and ardent sportswoman, who is said to have the remarkable record of fifty-two trout in fifty-four minutes."

Crosby was so inspired by the great outdoors that she wrote an article proclaiming the joys of fishing, signed it with her new nickname "Fly Rod," and sent it to the editor of the local newspaper, the *Phillips Phonograph.* He replied with the words every writer wants to hear: "That's mighty good stuff," he replied; "send some more right away!"

A gifted storyteller, Crosby was soon writing a popular outdoor column, "Fly Rod's Notebook," which eventually was syndicated in major newspapers as far away as Chicago, and writing pieces for national magazines like *Field & Stream* and *Shooting and Fishing,* and Maine Central Railroad's *Maine Central.*

She wrote tales of fishing and hunting, and reported on the people, wildlife, and adventures she experienced around the hotels and camps of Maine's outdoors, which were just beginning to be linked to the population centers of New York and Boston with improved railroad service. "Fly Rod told running stories from column to column and made allusions to earlier tales," wrote biographer Julia Hunter. "She kept her tone light and pleasant—the world she wrote about was run by excellent guides and affable sporting camp managers whose cooks set wonderful tables and served up as much of the guests' catch as was desired."

One day, she marched into the office of Payson Tucker, head of the Maine Central Railroad, with a business-building brainstorm. "If you want to give Maine and the Maine Central Railroad some advertising of permanent value, I know how you can do it," she declared. "Send me with a genuine Maine log cabin, a bunch of guides and a lot of stuffed deer and moose and birds to the [first annual] Sportsmen's Show in New York [at Madison Square Garden in 1895]. I'll do the rest." Until then, Maine was a distant second to New York's Adirondack Mountains region in the competition for vacationing sportsmen, fishermen, and hunters, but Crosby had a hunch that the new railroad lines to the lush wilderness of Maine could transform the state, if only people knew of the natural paradise that awaited them.

Tucker instantly knew a great idea when he heard it, and made a deal with Crosby on the spot. "You are a great young woman," he announced; "you will be the Maine representative in full charge [of the exhibit]." He hired Crosby to be Maine's first salaried, full-time publicity agent, to put together the New York exhibit and to put on similar exhibits at arenas around the Northeast. "Taking advantage of her talent for showmanship," wrote historian Austin Hogan, the railroad "put her in charge of their exhibits and with a coterie of guides and two beautiful Indian girls, Fly Rod pulled them in by the thousands."

Crosby's first big show, the spring 1895 Sportsmen's Exposition in New York City, was such a roaring success that she could hardly believe it. "Six to eight thousand attendance in the afternoon," she reported, "I never saw such a crowd in my life. You could not get near the Maine Central log cabin." Her team handed out three thousand flyers in a single day. Crosby cleverly promoted her fishing and hunting guides as "crack guides" who "will dress just as they do when on hunting and fishing trips, and at all times will be ready to tell hunting and fishing yarns." This brought potential customers face-to-face with the rugged, charming men who would be their companions in the woods, and it worked brilliantly. That summer, a record-breaking five thousand new tourists visited the state of Maine, and Crosby was given the credit.

Cornelia "Fly Rod" Crosby, America's first female fishing superstar. (Maine Historical Society)

For the next Sportsmen's Exposition at Madison Square Garden, in 1896, Crosby staged a bigger extravaganza. She even talked the federal government into sending the special-purpose railway fish tank "U.S. Fish Car No. 1" from Washington, D.C., all the way up to Maine to pick up one hundred trout and salmon, then ship them down to the show. The *New York Times* raved: “From the State of Maine there is to come to the exposition one of the most marvelous exhibits that was ever attempted in a show of this character. So effective was the display made by the people of the State a year ago in attracting sportsmen to the woods of Maine, that those interested in catering to the sporting visitors have combined to pay the expenses of a most complete exhibit, in the charge of Miss Cornelia T. Crosby, who is known to readers of sporting literature by her nom de plume of ‘Fly Rod.’ Miss Crosby is adept in the use of the rod and the reel, as well as an authority on the trout and salmon. In order to aid her exhibit, the [federal] Government, for the first time in the history of the Fish Commission, has placed at her disposal one of its fish cars, so that she might bring here a number of salmon and trout for exhibition. These have arrived safely and will be shown alive in the five glass tanks that have been provided.”

The new show was another smash hit, proving that Crosby wasn’t just a champion backwoodsman—she was a show business genius. As she gracefully demonstrated fly-casting into the fish tanks for the Madison Square Garden crowds, Crosby wasn’t just fishing, wrote one fishing historian: "she was toppling the shaky foundations of conventional arguments about a woman's proper 'sphere.'" She was blazing a trail for all women.

Crosby's triumph coincided with a new wave of interest in fishing among American women, powered by improved rail access to remote fishing spots and the availability of new generations of fishing gear. "By the end of the nineteenth century," wrote scholar David McMurray, "fisherwomen were able to be more active and aggressive in seeking out new opportunities for adventure within the North American wilderness as compared to the more sedentary demeanor of past generations of female anglers." He added, "Victorian women benefited from the mass production of fishing tackle which made angling accessible to those urbanites who had no experience in tying their own flies, making their own lines, or building their own rods."

Soon Crosby and her exhibits were credited with drawing thousands more new tourists and sportsmen to Maine. Hotel and camp owners and wilderness guides were overbooked—and overjoyed. Maine's economy boomed. Tourists poured in, from as far away as England and India, with lots of repeat customers coming from Boston, New York, and Philadelphia.

Heeding Crosby's call, thousands of women flocked to the Pine Tree State to camp, canoe, hunt, and fish.

Crosby staged more and more exhibits and promoted her state tirelessly, dubbing Maine "the Nation's Playground," whose "gold mine is the fish and game." She lobbied for wildlife management laws to "keep Maine what it is today—the best hunting and fishing preserve in America." In 1898, a local newspaper reported that Crosby's "prowess as a huntress and as a fisherwoman has been heralded all over the country. Her example has done much to encourage women to lead this kind of life. Her companionship is much sought by those desiring to emulate her skill with the rifle or rod. She thinks nothing of taking a forty-mile tramp, and has gone into the woods with the snow two feet deep on the ground. She can follow a trail with the sagacity of an old woodsman."

In 1897, the Maine Legislature honored Crosby by ceremonially issuing her the first official Maine Guide License (though technically she never worked as a guide herself) to honor the fact that she lobbied hard to create the revenue-generating program, which raised funds to protect fish and game.

The next year, tragedy struck. While getting off a train, Fly Rod suffered a devastating knee injury. The details aren't clear; she may have slipped on a piece of coal, or her skirt may have been dragged by the train's wheels. It disabled her for the rest of her life, and curtailed her fishing, hunting, and writing. Still, she managed to live on, despite failing health, until 1946, to the ripe age of ninety-two.

Fly Rod Crosby racked up a dizzying list of adventures and achievements. She championed the use of the lightweight fly rod and artificial lure for women, and wrote what may have been the first nationally syndicated outdoors column. As an adviser to the influential Maine Sportsmen's Fish and Game Association, she promoted wildlife conservation by fighting for responsible game management, catch-and-release fishing, bag limits on deer, salmon, and trout, and expanded fish stocking programs in lakes and ponds.

Crosby was the first woman to get a fishing license in Maine, she helped establish the first boy's camp in Maine, she was the first Maine woman to join the New England Women's Press Association, and she helped establish the Maine State Museum. Her many friendships included the famed sharpshooter Annie Oakley, which led historians Julia Hunter and Earle Shettleworth Jr. to note, "The two women had much in common as professional athletes eighty years before professional women's sports would even begin to be taken seriously in North America." Today, there is a beautiful wilderness trail in western Maine named in honor of her, The Fly Rod Crosby Trail.

The *New York Times* reported that Crosby was a "much sought after" speaker who could entertain "a host of sportsmen" with her storytelling, and also handle the "inquiries of ladies who, like her, have a fondness for fishing and hunting."

"How I wish more of the ladies would leave their party dresses at home," she wrote in a 1901 *Field & Stream* article, "take the short skirt, the sensible boot and come to Maine. I'll teach you to fish and tell you where to enjoy life."

In 1908, she wrote, "Why should not a woman do her share of fishing, hunting, tramping and mountain climbing and ask no odds of the men? There is no more graceful, healthful and fascinating accomplishment for a lady than fly fishing, and there is no reason why a lady should not in every respect rival a gentleman in the gentle art."

Through her celebrity appearances, shows, and syndicated columns, Crosby turned Maine into a fishing capital, taught thousands of men, women, and children about fly-fishing and wildlife conservation—and symbolically opened the doors wide for millions of

American women to savor the great sport of wilderness fishing.

Women have been key players in the sport of fishing ever since the dawn of the American nation. As early as 1760, a British traveler reported visiting a fully coed fishing club in Philadelphia, complete with full fishing gear, uniforms, and dance parties. "There is a society of sixteen ladies," he wrote, "and as many gentlemen, called the Fishing Company, who meet once a fortnight upon the Schuylkill. They have a very pleasant room erected in a romantic situation upon the banks of that river, where they generally dine and drink tea." He continued: "There are boats and fishing tackle of all sorts, and the company divert themselves with walking, fishing, going up the water, dancing, singing, conversing, or just as they please. The ladies wear a uniform, and appear with great ease at the neatness and simplicity of it. The first and most distinguished people of the colony are of this society; and it is very advantageous to a stranger to be introduced to it, as he hereby gets acquainted with the best and most respectable company in Philadelphia."

In the early 1800s, a number of American women enjoyed fishing as a pastime, and frontier women pitched in to fish in lakes and streams for food for the family dinner plate. But fishing was largely considered to be a man's affair, in line with the sharply defined social position of women in society as mothers and homemakers—and not much else. Culturally, however, the nation was gradually and profoundly transformed by the women's rights movement that began at the 1848 Seneca Falls Convention, and intensified through the turn of the twentieth century until 1920, when women nationally won the right to vote.

In the wake of the Civil War and the rise of vast new American settlements and cities, people were eager to enjoy nature and to explore the new lands of the West. Women were no exception.

By the second half of the nineteenth century, millions of women were entering the workforce as factory, agricultural, and service workers, and some were entering business and the professions. They were just as interested in leisure activities as men were, and just as attracted to wilderness, nature, and sport, but the question was how to channel these passions in a society that historically considered women to be fragile and delicate creatures. The answer, for many women, was angling. According to scholar David McMurray, "angling was the preferred outdoor pastime of many middle- and upper-class, nineteenth-century North American women who had inherited it as a respectable sport from their Early Modern predecessors [European women in the 1400s through 1700s]. Evidence for this comes from the examination of diaries, pictures, photographs, and literature that demonstrates a long tradition of women in angling from the Renaissance through to the Victorian era."

This mass movement of women to the outdoors and to fishing was driven by women themselves, but a number of men championed the cause as well, often in the new sporting media that appeared after the Civil War. In its first issue in 1873, *Forest and Stream* magazine, the predecessor of today's *Field & Stream,* editor Charles Hallock stated, "Ladies are especially invited to use our columns, which will be prepared with careful reference to their personal perusal and instruction." Soon a father wrote in to report that "girls everywhere are learning the use of the rod." He encouraged sportsmen dads to take their daughters out into nature, to cherish and praise their exploits, and to appreciate the fact that "nature is ready to give impartially to boy and girl alike."

In subsequent issues, *Forest and Stream* editors and writers, male and female, asserted that "now it is no unusual thing to see a woman fishing a stream, following the dogs, or sailing a yacht," there was "no good reason why there should not be as many and as good

Young American angler Doris Gottscho, circa 1919.
(Library of Congress)

sportswomen as there are sportsmen," and "an increasing number of women were "accompanying their husbands or brothers on fishing or shooting excursions." According to another writer, "A woman could ride horseback, shoot, row, climb mountains and fish with the same ease and proficiency that she can preside at a social function or indulge in a literary lucubration." By encouraging women to enjoy outdoor sports like fishing, magazines were also attracting advertising from

An American fishing scene, circa 1906. (Library of Congress)

fishing rod and tackle manufacturers, who realized that women anglers and their families were a big new consumer market.

By 1889, American women were taking so much to the outdoors that writer Emily Thackvay devoted an *Outing* magazine feature piece to the trend, in which she noted, "The most encouraging signs of the times in American life is the increasing love for out of door life and sports, indulged in now not by men alone, but also by those who used to be termed 'the weaker sex.'" She added, "No woman is so free as an American girl; if she uses up her freedom in summer to build up her bodily strength and refresh her mental vision, if she tramps through the forests and over mountain peaks, drinking in ozone and beauty with every breath, if she elects to camp out under the starlit canopy, paddle her slender canoe and cast her trout fly upon the unsullied mountain lake, gathering there with the silver-white lilies that star its surface the 'seed of white thoughts, the lilies of the mind,' she will come home in the late autumn with rosy, nut-brown cheeks and a fresh spring in her steps."

On August 14, 1898, a *New York Times* headline declared: "Feminine Izaak Waltons: There Are Many of Them Who Enjoy the Delicate Sport of Angling." An unnamed female angler explained in the article how women could be superior to men in fishing skills: "I have fished with men who enjoy fishing, and who can endure a great deal of hardship, but they say I can endure more hardship than they. I do not wade in the water, but I walk from five to twelve miles, and the men are tired and I am not. And I have had luck in one way in getting the biggest fish—not the quantity, but the quality. But that is because I have patience. Women can do better in that way, for they have more patience. A man will throw his fly and wait a little while, and then he says the fish will not bite, and he goes away, when perhaps if he waits a second longer, he gets the fish." A woman's patience and endurance, she felt, made for better angling.

What did women discover as anglers? Many of the same things that men did—adventure, relaxation, the beauty of nature, physical joy, and spiritual comfort. But they also used fishing to channel their dreams of escape, mobility, independence, and achievement in an age when their social roles were evolving radically.

One woman of this era, Sara Jane McBride, turned her passion for fishing into a career. Born in 1844, she learned fishing and fly-tying from her father, famed fisherman and tackle seller John McBride. As a child, Sara wandered the creeks and rivers around their home in upstate New York, closely examining water insects and fish. Soon, tutored by her dad, she was tying flies of her own, and by the mid-1870s, she took over the family business and opened up a store in New York City. One of her most popular fancy fly patterns was the Tomah Joe, inspired by the work of a Native American fly tyer. In 1876 she won a bronze medal for her designs at the Centennial Exhibition in Philadelphia.

Also in 1876, Sara published articles on "The Metaphysics of Fly Fishing" in leading outdoor journal *Forest and Stream,* which hailed her as a rising star in the fishing industry, and an authority on entomology and fly-tying: "Miss McBride's skill in fly tying has long been known to the angling fraternity. For neatness, finish and skill, the flies made by Miss McBride have but few equals. Having removed from Mumford to New York City, this lady has now opened an establishment at 889 Broadway, where flies adapted to all seasons or localities may be found. Amateurs desirous of making their own flies can be supplied with all the materials necessary. The general fishing public can find a full selection of rods, reels, lines, with all the newest trolling baits. In her present locality it is to be trusted that Miss McBride will not only retain her former extensive patronage, but will be able to secure a large portion of new business."

Mysteriously, in 1879, she abruptly closed up her New York City shop, moved back to upstate New York, and vanished from history.

Nobody knows what happened to Sara McBride, the first professional woman fly tyer in America. But what's certain is that she was a pioneer woman entrepreneur, whose skill, independence, and love of fishing blazed an early symbolic trail for legions of anglers who followed.

Another woman who used fishing to become a pioneer of business was Mary Orvis Marbury of Manchester, New Hampshire. Like "Fly Rod" Crosby and Sara McBride, she achieved greatness in the world of fishing with her writing, business, and technical skills.

Her father, Charles F. Orvis, founded the Orvis fishing tackle and sporting goods company in 1856, the same year she was born. The business consisted of a shop selling fly-fishing rods and flies made by the company, and branched out into a booming mail-order catalog business. When she was twenty years old, Mary took over the company's commercial fly production department, supervising an all-female staff of up to ten workers. Also in 1876, her first year in the business, she won first prize for her exhibit of flies at the Centennial Exhibition in Philadelphia.

Mary Orvis Marbury, who helped build a fishing business empire. (Cornell University Library)

For twenty-eight years, Mary devoted her life to the art of fly-tying and became America's leading expert on the critical relationship between fish and their main meal, bugs. By 1890, some 434 different fly patterns were featured in the Orvis catalog, all produced by Mary's team of fly tyers.

When Mary's father sent hundreds of letters to customers and anglers across America asking for in-

Fly designs by Mary Orvis Marbury. (Cornell University)

formation on local fishing patterns, she compiled the responses along with her own research into a first-of-its-kind 1892 book that became an instant fishing classic and bestseller, *Favorite Flies and Their Histories.* In the book, she wrote of the "fresh vigor and strength" and "excitement without worriment" obtained by fishing in the great outdoors, and told stories of the history and development of different flies, like "No. 192," which was named after a beloved fisherman-educator: "the story of the fly is, that one time, when this famous angler was fishing, he ran short of flies, and, to create something of a flylike appearance, he fastened the petals of buttercups on his hook, adding bits of leaves or grass to imitate the wings of a fly. This arrangement was so successful that it led to the making of the fly with a yellow silk body, since then so widely known as the Professor." The book, which was reprinted nine times in the next four years, featured color illustrations and details of 291 patterns from thirty-six states, two territories, and Canada, and became the bible of fly information for American fishermen.

Mary Orvis Marbury's personal life was not a happy one—she had a brief marriage and a son who died young, but her professional achievements were substantial. She was the first person to organize American fly patterns, and she helped build the foundations of a great fishing, hunting, and sporting goods business, the Orvis Company, which endures today as a dominant force in the field.

I can relate to this in my own life. The vital role of women in inspiring and building an outdoor products business is reflected in my own company, Duck Commander. It never would have become what it is without the women of our family.

My mother, Kay, in full partnership with my father, ran the business, paid the bills, and ordered the inventory, and then that passed on to the other women in the family. My wife, Korie (an enthusiastic recreational angler, by the way), is very instrumental in the business today, and early on she worked on the business side of our TV show more than I did. Now she runs our licensing and design business. She is involved in pretty much every business that we make. My sister-in-law Missy has been involved in the business, too, right from the start. As so many other Americans do, we all work together as partners—and as family.

We are glad to know that there are some sportsman who consider their wives or sisters as real chums on a trouting excursion, instead of leaving them at home wishing "that they were men" that they, too, might enjoy an outing and leave their perplexing cares at home on a shelf. A woman will soon learn to love the ripple or roar of a trout stream and the song of the reel as much as her husband or big brother, and look forward expectantly to a "day off" in the mountains enticing the wary trout.

—Mrs. F. Cauthorn, *Forest and Stream*, 1892

The first fish I caught on a fly rod was magic. It was the most exhilarating experience. And after I released that fish back into the water, that was it. I was addicted. . . . For me, the moment just before a fish takes a fly, just when he commits, is the most exciting in fly fishing. It is so exhilarating, and I become so focused. All of my daydreams end, life's minutiae melts away, and I'm in the moment. I do fish a lot in solitude, but I really enjoy fishing with my friends. And a lot of my really close friends began as strangers . . . but through some conversation somehow we stumbled upon the fact that we really share this passion, and it brought us together, and now they're wonderful friends.

—Diana Rudolph, Angler

As soon as the sport of saltwater fishing took off in the 1890s, American women made their mark on it. Many of them went to sea with their husbands to enjoy the pursuits of the new wealthy class that emerged in America's "Gilded Age," and these women soon proved their skill as independent forces to be reckoned with on the stage of sportfishing.

In May 1891, while on a fishing trip with her husband on Florida's Caloosahatchee River, experienced angler Mrs. George T. Stagg hooked and landed a world-record 205-pound, 7-foot, 3-inch tarpon, after

Famed Western novelist Zane Grey was also one of America's bestselling nonfiction writers on fishing.

what *Forest and Stream* reported was a "gallant fight of one hour and twenty-five minutes." The prize catch was put on display at the 1893 World's Fair in Chicago.

A dispute over a saltwater catch triggered a controversy that involved famed writer and fisherman Zane Grey, then the world's bestselling writer. In 1920, he landed a season-record 418-pound broadbill swordfish, which was then the most-sought-after big game fish. According to one account, the pugnacious Grey "quickly became a nuisance around the [Tuna Club of Avalon, on Catalina Island, California, then the center of the angling world] clubhouse as he recounted, over and over again, the minutest detail of his fight." Grey became so obnoxious about his catch "that most Tuna Club members tried to avoid him, ducking down behind newspapers when he entered a room, or excusing themselves to run an errand in town when he sat down on the arms of their chairs."

The next summer, Mrs. Keith Spalding, the diminutive wife of the new club president, who weighed less than one hundred pounds and stood under five feet tall, caught a bigger, 426-pound broadbill. Zane Grey's fellow club members teased him relentlessly about it, to the point that he was heard angrily declaring that she never could have done it herself, someone must have helped her. The club management demanded that he apologize in writing for the outburst—or resign. He did both.

It was the toughest fight that Sara Chisholm "Chisie" Farrington ever had with a fish.

It happened one morning in 1936, off Nova Scotia, when she hooked a 493-pound tuna. A titanic struggle ensued, a battle that lasted a full 10 hours and 25 minutes.

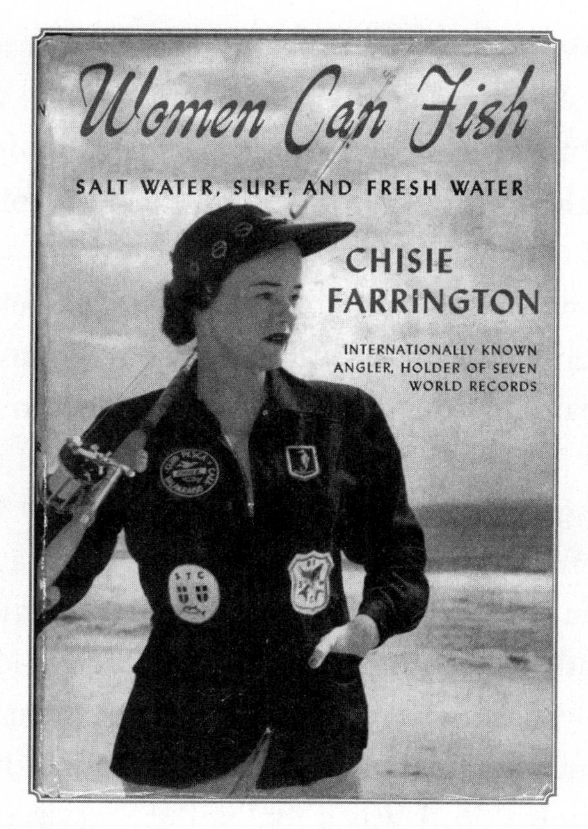

The cover of Sara "Chisie" Farrington's 1951 account of her legendary fishing exploits.

Halfway through the grueling physical contest, Farrington paused to wonder what on earth she was doing. She was a highly experienced angler, one of the best, in fact, but she recalled, "At this point I thought I was crazy. What would I have, if I landed this fish? Nothing but a cold, tired, dead fish that was not nearly as large as the one I had caught two days ago.

Then why didn't I give up? I guess there was a certain amount of pride. I thought of all my friends who had built up my fishing and I didn't want to poop out on them. I thought to myself, 'I will keep on going until I faint.'"

Soon, however, Farrington faced total exhaustion. "I can't keep going on forever, eventually I must faint," she despaired. "But that cold air blowing in my face wouldn't let me faint," she remembered, "so I just kept on reeling the skiff up to the tuna and then the tuna would pull out more line. I could do nothing at all with him; then why didn't I give up? I thought of all the letters I could write, the Red Cross work I could do, the things I could accomplish in the hours spent battling this damn tuna. Then why didn't I give up? It was no crime to give up. I was a fool not to, but the longer I kept at it the madder I got, and the madder I got the more determined I got to boat the tuna if it [was] the last thing I ever did!"

Incredibly, after ten and a half hours, when Farrington's tuna was hoisted up on the dock for a picture, it had a companion—another giant tuna, this one a 720-pounder that she had captured two days earlier, in just 1 hour and 38 minutes.

Raised in elegant circles in Manhattan and Long Island, Farrington attended the exclusive Spence and

Miss Porter's schools, and in 1934 she married Selwyn "Kip" Farrington Jr., an accomplished writer and outdoorsman and author. She quickly developed a passion for saltwater fishing, like her husband, despite her weak right leg and hand caused by childhood polio. Their honeymoon, naturally, featured fish. "Kip suggested we go chumming on our wedding trip," Farrington remembered. "This sounded very romantic to me, but I hadn't caught on that it would mean grinding up mackerel off the Portland Lightship at 4am in a very rough sea." Her husband's wedding gift to her was a fishing harness.

In the 1930s and 1940s, the couple romped around the world on fishing voyages, which Mrs. Farrington wrote about in articles for *Harper's Bazaar, Vogue,* and *Mademoiselle,* and in a 1951 book called *Women Can Fish.*

Farrington racked up an impressive list of accomplishments in her sport, including eleven world saltwater records as measured by the International Game Fish Association (IGFA). She was the first woman to catch a giant bluefin tuna on rod and reel (Nova Scotia, 1935), the first woman to catch two marlins in one day (off South America, 1939, considered one of the greatest possible saltwater feats), and was the first angler to catch a tuna in the fishing grounds off Rhode Island.

Farrington was so good that she earned the admiration of the Great Fisherman himself, Ernest Hemingway, which was no easy feat. When she became the first woman known to land a broadbill swordfish off the South American coast, which weighed 584 pounds, Hemingway dashed off a congratulatory cable to her: "Perfection. The real record is to take the first one, because if you catch the biggest fish, someone eventually is going to catch a bigger one."

To have Papa call you "perfect," well, that was the ultimate compliment to an angler.

As a world-champion, globe-trotting angler, Chisie Farrington came of age in the mid-twentieth century, an era when American women were increasingly asserting themselves as partners, competitors, and collaborators in many fields of professional and sporting life. Women had mobility, education, spending power, and mass media to cater to their interests, and, following the early example of independent women like Sara McBride, Mary Orvis Marbury, and Cornelia Crosby, some women were ready to follow their passion for sport to the ends of the earth—and completely on their own terms.

One day in 1981, deep in the Amazon rain forest, a tall sixty-one-year-old woman in a sun hat jumped out of a dugout canoe and splashed ashore.

She stood on the banks of Brazil's Rio Branco River, forty miles south of the Equator. It was one of the most dangerous spots on the planet, teeming with electric eels, vampire bats, man-crushing snakes, piranhas, wild pigs, jaguars, and venomous attack insects.

"Splash!" she shouted to a friend wading ashore. "Kick your feet up!"

"Why?" the man asked.

"Stingrays!" she said.

After the man scampered ashore, the woman explained how in the Amazon, stingrays lay in warm, shallow water like this, and when you stepped on one and it bit you, you could look forward to twelve hours of such agonizing pain, you would thrash so hard you'd break your own bones. If you splashed ashore instead, the stingrays would slither aside.

The woman broke out her gear and started doing the one thing she loved doing more than anything else in life: she went fly-fishing. During this trip she hoped to catch specimens of exotic fish species like tucunarè, arawana, trahira, pirapacu, and arapaima.

"People think that I'm a mad, brave old woman to come out fishing in a place like this, living like an Amerindian," she said. "But, hell, I've been robbed twice on the streets of D.C., and each time it was more frightening than anything I've met in a rain forest."

Her name was Kay Brodney. She was the Indiana Jones of fishing.

In her day job she was department head of the Life Sciences Subject Catalog Section of the Library of Congress. In her other life, she traveled the world in search of exotic fishing adventures, from California and Florida to Nova Scotia, Argentina, and Brazil.

Brodney grew up in Fond du Lac, Wisconsin, where she learned to fish for bass and pike. In her twenties, as a self-described "classic dropout," she hitchhiked to California and settled there. One day in 1948, by chance, she came across a casting tournament in San Francisco's Golden Gate Park. She loved what she saw. The sight of dozens of competitors whipping their fly rods around in the splendor of that park confirmed her destiny—she would spend the rest of her life fishing as much as she possibly could. "I saw those lines swishing about and it changed my whole life," she recalled. "Women weren't recognized for doing much distance casting then, just accuracy events. Once I took third place in the Western Championship down at Long Beach. I went to get my prize but I found I hadn't qualified because I wasn't a man."

Through her mid-thirties, she went through fifty different jobs, including waitress and railroad clerk, all to finance her fishing habit. She bought a Volkswagen

Camper and drove all over northern California angling for shad and steelhead. In Seattle, she went to work for the state fishery service and earned a B.A. in zoology. She saved up a thousand dollars, which she spent on an unsuccessful tarpon-fishing expedition to Baja California. To continue financing her fishing, Kay earned a master's degree in library science at Rutgers University in New Jersey. "Then I got a job in the Library of Congress," she explained in 1981. "I've been there 16 years, and I can blow five or six grand a year on fishing!" She hand-made her own fly rods, lines, and flies. In her professional and sporting lives, Brodney was emblematic of a modern American woman: adventurous, highly educated, and intensely self-directed.

"She did it all," reported her friend and fellow globe-trotting fisherman Jim Chapralis. "Kay fished in Argentina and was one of the first anglers to take a dorado on a fly. She fished the jungles of Brazil camping out in areas where the Indians were considered unfriendly." Kay explained to him, "We didn't carry any weapons, and I think they knew this and therefore we weren't considered threats." What if the natives attacked? "If we die, we die," she said. In 1962, Kay landed a giant 137.6-pound tarpon on a 12-pound tippet in the Florida Keys, with baseball-and-angling legend Ted Williams looking on.

One day, while sitting in a stump in the Amazon jungle, Kay noticed that a tiny insect had bitten from her feet up to her waist. At this moment, she apparently contracted a disease that was never properly diagnosed, and she suffered greatly from it for the rest of her life. Even when she was too weak to cast her fly rod, though, she enjoyed sitting in the boat while her friends fished and cheered her up. Kay Brodney died in 1994. When it came to fishing, she had truly done it all.

Today, a living legend casts with her fly rod on the streams of New York's Catskill Mountains, teaching new generations the joys of fishing.

In the person of this one woman, you can trace much of the historical trajectory of American women fishing at the pinnacle of the sport, over more than seven decades. Inspired by her father, she fell in love with the sport as a girl, pursued her passion so intensely that she developed a career out of it, eventually became one of the best in the world at her craft, entered into a long marriage and business partnership with an equally talented fisherman, and continues to this day as a successful fishing entrepreneur.

She is Joan Salvato Wulff, the ninety-year-old First Lady of Fly Fishing, the greatest female fly caster in America, and a twentieth-century pioneer and master

Joan Wulff at thirteen, already on the path to fishing greatness. (American Museum of Fly Fishing)

teacher of the sport. "She has done for casting what Stephen Hawking did for physics," declared *Fly Rod & Reel* magazine when it named her Angler of the Year in 1994. "She stands alone in her ability to communicate its mechanics as well as its human aspects."

Joan's father owned a rod and gun shop in New Jersey. When she was ten years old she borrowed his fly rod without permission and tried to cast in a local pond, quickly losing the tip. Her dad figured it was time he taught her how to fish, and she's been fishing ever since. She was amazed by how beautiful a feeling it was. "I loved fly casting because of the grace and beauty of it. That felt like what I was meant to do. And I think it was. I was born to fly cast."

"As is true of many of the women of my generation who fly fish, my father was a fly fisherman," she explained. "Mother was not. It became apparent the first time I accompanied them for an evening of fly fishing for bass that Dad had all the fun while Mom got yelled at for not keeping the rowboat at the right distance from bass cover. Unencumbered by the knowledge that women didn't fish, it was obvious to me then, at age five or six, that it was better to be the fisherman than the rower."

By age eleven, Joan had won her first casting competition title. By age thirty-four, she had won seventeen

Joan Salvato Wulff, the first lady and master teacher of fly fishing. (R. L. Winston Rod Company)

national casting titles and an international baitcasting title in London, and set a woman's record distance cast of 161 feet. In 1951 she became the first woman to win the National Fisherman's Distance Fly Championship, with a cast of 131 feet versus an all-male field of competitors. In 1952, she turned professional. She went on the road to promote fishing products for sponsors, and audiences loved to watch her conduct trick-casting demonstrations in an evening gown and high heels.

In 1967 Joan married legendary fisherman Lee Wulff, and together they established the Wulff School of Fly Fishing, on the Beaverkill River in Lew Beach, New York; it endures to this day. She wrote fishing columns and books, helped to found the Catskill Fly

Fishing Center and Museum, made multiple television appearances on ABC, ESPN, and PBS, and, as an instructor, has inspired countless women, men, and children to take up and master fishing, including her grandchildren.

"Why do I love fly fishing?" she once asked. "I love this sport because of its depth. There's art. And literature. And life sciences. And craft. But I really love it because of the wild and beautiful places it takes me, where I can connect to another of God's creatures and feel its life force through my fly rod, and release it."

Today, women are the fastest-growing segment of the American fishing industry. They number nearly 9 million anglers, and they spend more than $10 billion on fishing gear, trips, and fees every year.

8
Fishermen in the Modern White House

Fish are constantly doing the most mysterious and startling things; and no one has yet been wise enough to explain their ways or account for their conduct.
The best fishermen do not attempt it; they move and strive in the atmosphere of mystery and uncertainty. In these circumstances fishermen necessarily see and do wonderful things.

—Grover Cleveland, twenty-second and twenty-fourth President of the United States

On April 28, 1937, President Franklin D. Roosevelt decided to escape from the White House. He had to get out of town.

Things were heating up for him in Washington, both weather-wise and politically. The Great Depression was still raging. Roosevelt was locked in full-scale combat with much of Capitol Hill over Congress's move to overturn some of FDR's beloved New Deal programs, and the president was pushing a highly controversial, and doomed, plan to expand and "pack" the Supreme Court in his favor. A typically brutal hot summer would soon strangle the nation's capital.

Roosevelt ordered his bags packed, and he set off for Louisiana. He wanted to get down to the Gulf Coast and do some fishing.

When FDR had been invited to go on a fishing trip to the Gulf of Mexico, he practically leaped at the opportunity to relax by enjoying one of his favorite pastimes. Despite his polio-induced paralysis below the waist, a disability that he chose to conceal in public and that confined him to a wheelchair, FDR was a lifelong fisherman, amateur sailor, outdoorsman, and nature lover. He craved the invigorating air and sunshine of the ocean, and the relaxed company of friends on long, lazy fishing vacations.

That day, April 28, the presidential party left Washington by train, and when they arrived in New Orleans, FDR took his aides out to Antoine's restaurant in the French Quarter, one of the finest dining establishments in town, where they savored the famous *Pompano en Papillote,* a tantalizing shellfish-and-white-cream-sauce dinner said to equal anything available in Paris.

It was a delightful seafood meal, especially for a man who lived in a prison of lousy food. At the White House, First Lady Eleanor Roosevelt had condemned her husband and all their guests to a culinary dark age that lasted all twelve years of their occupancy. For unknown reasons, Mrs. Roosevelt insisted that the White House kitchen would be run by a stern, culinarily untrained Hyde Park housekeeper and neighbor named Henrietta Nesbitt.

President Roosevelt yearned for the kind of high-end epicurean diet he'd grown accustomed to as a New York aristocrat, including seafood delicacies like lobster and caviar, but FDR evidently never confronted his wife about Nesbitt's offerings. Ernest Hemingway recalled one of Nesbitt's White House dinners as the worst meal he'd ever had, consisting of "rainwater soup, rubber squab [pigeon], a nice wilted salad and a cake some admirer had sent in." A White House maid reminisced that Nesbitt "stood over the cooks, making sure that each dish was overcooked or undercooked or ruined one way or another." The word quickly went out among Washington insiders: whatever you do, grab something to eat before you go to the White House.

Franklin D. Roosevelt about to dig into a world-class seafood dinner at Antoine's of New Orleans, joined by his son Elliott (far left), on the eve of his tarpon-hunting expedition to the Gulf of Mexico, 1937. (courtesy of Antoine's of New Orleans)

But at Antoine's of New Orleans, free of Nesbitt's torment, and between jauntily savored puffs from an endless stream of unfiltered Camel cigarettes twisted into his cigarette holder, President Franklin Roosevelt enjoyed an expertly cooked version of pompano, a fish that the *Washington Post* once called "one of the most highly prized Gulf Coast fish," with meat that "is firm, sweet and sublime."

It was the perfect start to a presidential fishing trip.

FDR's housekeeper and kitchen supervisor, Henrietta Nesbitt, who presided over what some considered to be the worst White House food of the modern era. (Library of Congress)

The next day, the president's party boarded the Navy destroyer USS *Moffett* and sailed for the Gulf, with the USS *Decatur* providing an escort screen. On May 1, the ships rendezvoused with the presidential yacht *Potomac* in the Aransas Pass channel. The luggage was transferred to the yacht, and FDR was transferred to a small launch so he could start fishing. FDR's movements between boats, ships, trains, cars,

President Franklin D. Roosevelt leaves the USS Target *in March 1935 at the end of a fishing trip, in Jacksonville, Florida. (Library of Congress)*

and buildings were always a complicated and secret affair, given his physical condition. Out of view of the public and reporters, strong men had to hoist him up and carry him in their arms during the transfers, and special pulleys, elevators, hoists, and ramps had to be prepositioned wherever he went.

Through all the behind-the-scenes fuss, Roosevelt smiled, laughed, and never seemed to mind, especially when he got to go fishing, which he tried to do as often as possible. He loved fishing with a passion, especially in the ocean. In October 1935, aboard the *Houston* in the Cocos Islands, he wrote a giddy letter to his mother: "We are having a most delightful cruise, and at this charming spot the fishing is excellent. I have caught two very large sailfish—one on Wed.—110 pounds. And one today—134 pounds. The result is that my muscles are rather sore, but it is good for my figure and I get lots of sleep and sun and fresh air." After a two-week deep-sea Caribbean fishing cruise aboard millionaire Vincent Astor's luxury yacht, Roosevelt wrote wistfully, "We hugged azure skies, golden sands, turquoise depths, lush pampas, intriguing inlets, basking lizards, swooping seagulls, winking stars, snapping turtles, lovely doves, verdant seaweeds, and perfect serenity."

Like many of his fellow fisherman-presidents, and rank-and-file fishermen as well, Roosevelt saw the main benefit of fishing as being not so much the catching of the fish as the mental relaxation it offered. "The objective of these trips, you know, is not fishing," he told a group of reporters crowded around his Oval Office desk on May 13, 1937. "I don't give a continental damn whether I catch a fish or not. The chief objective is to get a perspective on the scene, which I cannot get in Washington any more than any of you boys can. You have to go a long ways off so as to see things in their true perspective; because if you sit in one place, right in the middle of the woods, the little incidents that don't mean a hill of beans get magnified by a President just as they do by a correspondent."

By now, American life had become much more urban and suburban, and fewer Americans lived in nature or very close to it. The frontier was closed. Modern living imposed new stresses and frustrations on Americans, and fishing offered a wonderful chance to escape—especially for the most powerful man in the world, the President of the United States.

Franklin Roosevelt was one of those rare amateur fishermen who had both great natural skill and consistent good luck. "He could fish from the back of pleasure boats and from specially constructed seats on naval vessels," wrote presidential fishing historian Bill Mares. "Indeed, Roosevelt became an omnivorous saltwater angler. He would fish for tarpon, sailfish, shark, or dogfish." According to journalist Robert Cross, "Roosevelt was a first-class fisherman, who usually was successful at netting a good catch even when his colleagues were not. He knew how to handle a rod and reel, and seemed to have a sixth sense as to where the fish would be biting. He also had patience, and would sit for hours in a small launch, trolling and simply wanting to hook a big one. While many deep sea fishermen used a leather harness to provide additional leverage for them and their rods, Roosevelt did not need one because of his powerful shoulder and arm muscles."

Roosevelt brought back boxloads of trophy fish and angling mementos from his trips, and displayed them in a waiting room across from the Oval Office. He added an aquarium and dubbed it the "Fish Room."

On day one of his 1937 Gulf Coast escape, FDR got right down to fishing, and quickly caught a king mackerel. A Navy seaplane delivered the White House mail pouch. Roosevelt worked through his paperwork, had dinner, then fished off the quarterdeck of the *Potomac* until twilight. Searching for his career-first tarpon, Roosevelt felt four solid tugs, but landed nothing.

The next day, Monday, the president snagged a tarpon and started working it in. When guide Barney Farley saw the fish, his heart sank. "The hook had torn a large hole in the tarpon's mouth," remembered Farley. "I knew that when I took hold of the leader, the tarpon would jump and the hook would fall out or tear out. I told Mr. Roosevelt I was about to experience the most embarrassing moment of my life."

He told FDR, "I am going to lose your first tarpon." The fish shook off the hook and disappeared. "You called that right!" said FDR. Roosevelt finally landed a four-foot-plus tarpon, and proudly held it up for nearby photographers.

On Friday morning, FDR was handed a bulletin that the German zeppelin *Hindenburg* had exploded while attempting to land at Lakehurst, New Jersey, killing thirty-six passengers and crew members. Roosevelt wired a note of condolence to German Chancellor Adolf Hitler. Later that day Roosevelt's boat approached St. Joseph's Island to have a spontaneous lunch with political supporter Sid Richardson.

"When we got there we were faced with the problem of how to get the president in his wheelchair from the boat to the island," recalled guide Barney Farley. "Sid didn't have any way to unload him. He didn't have a dock for that. But he did have a cattle chute. We pulled alongside the chute and Sid explained he was going to roll him down."

"What in the world?" called out Roosevelt, "Do you mean you're going to roll me down that bull chute?"

"Mr. President," said Richardson, "you're the biggest bull that ever went down that chute!"

Franklin Roosevelt grew up in the lap of luxury, as the heir to a Dutchess County, New York, family of Dutch blue-bloods who had plenty of time to enjoy swimming, sailing, transatlantic ocean travel, golf, bird-watching, and casual fishing, which he learned from his father as a boy. As a young man, FDR learned to fly-fish, and within a year of his devastating polio attack at the age of thirty-nine, he was back fishing. At his health retreat and spa for polio victims at Warm Springs, Georgia, he enjoyed fishing for perch in nearby ponds.

Soon after becoming president in 1933, Roosevelt commandeered the U.S. Navy's cruiser USS *Houston* as his long-range presidential travel vessel, as well as his personal fishing vacation and party boat. On four occasions, in 1934, 1935, 1938, and 1939, the vessel was customized to handle the president's wheelchair and White House staff. In 1934, the *Houston* took FDR on a twelve-thousand-mile round-trip from Annapolis, Maryland, to Hawaii. On a twenty-four-day fishing cruise from San Francisco to the Galapagos Islands in 1938, Roosevelt and scientists from the Smithsonian Institution brought back eighty species of fish, including thirty new ones. Fishing from a whaleboat, Roosevelt caught several record-size fish and a 230-pound shark. According to naval historian James D. Hornfischer, Roosevelt was beloved by the crew and spent much of that twenty-four-day trip telling jokes and fishing on the ship's motor launch.

World War II greatly cut back on Roosevelt's fishing trips, but he loved escaping to the new presiden-

FDR and Winston Churchill enjoy fishing near Camp David, 1943. (FDR Library)

tial retreat he dubbed "Shangri-La" (later called Camp David, after Dwight Eisenhower's grandson) in Maryland's Catoctin mountains, and fishing in the nearby streams like Hunting Creek, which National Park Service employees kept heavily stocked with brook trout for the chief executive's fly-fishing pleasure.

One weekend in 1943, in the midst of the global conflict, FDR invited Winston Churchill to go fishing with him. Churchill remembered the scene at Hunting Creek: "On Sunday the President wanted to fish in a stream which flowed through lovely woods. He was placed with great care by the side of a pool, and sought to entice the nimble and wily fish. I tried for some time myself at other spots. No fish were caught, but he seemed to enjoy it very much, and was in great spirits for the rest of the day." Nearby, FDR's assistant William Rigdon looked on, and later remembered the scene: "The two men sat side by side on portable canvas chairs—the President pole fishing and Churchill smoking. The cigars created enough of a screen to protect both of them from mosquitos. They would sit there and talk by the hour."

The sport of fishing brought countless hours of peace and relaxation to Franklin D. Roosevelt, who desperately needed both as he grappled with two of the most titanic crises ever faced by any president—the Great Depression and World War II. By showing off glimpses of his fishing adventures to the public through newspaper photos and newsreels, he helped set the stage for the huge postwar expansion of sportfishing in America. Franklin Roosevelt's most lasting legacy to fishing were his Works Progress Administration (WPA) of the New Deal and the creation of the Tennessee Valley Authority (TVA), both in 1933, which tamed rivers, built dams and reservoirs, and established water management and river navigation programs. These initiatives triggered the detonation of game fish populations, which helped the sportfishing industry take off.

FDR was only sixty-three when he died in April 1945, probably from the cumulative effects of stress, heart trouble, lack of exercise, incessant smoking—and, perhaps, twelve years of Henrietta Nesbitt's relentlessly bad food.

Roosevelt was a famously compassionate man, who tried to comfort destitute farmers, unemployed workers, and wounded veterans alike during his epic twelve years in office.

He even had feelings for fish.

At one point during Roosevelt's 1937 fishing trip to the Gulf Coast, recalled his guide, the president fought a tarpon "over a distance of three and one half miles and one hour and twenty minutes" before the creature was subdued.

Roosevelt reached over, patted the conquered fish, and spoke to it in a spirit of chivalry.

"Thanks, old fellow, you put up a good fight."

Fish frolic through the pages of American presidential history.

FDR proudly shows off a catch from his epic fishing trip aboard the Houston, *1938. (Otto Schwartz, US Navy)*

Sometimes they're bit players. Other times they vanish for long periods. But occasionally, as in the case of several fish-loving chief executives, they appear practically as costars. "In rippling mountain brooks and lazy southern ponds," wrote journalist Lawrence Knutson, "in salt water and fresh, in hip boots and crisp three-piece suits, presidents have gone fishing since George Washington became the First Angler."

As we've seen, George Washington owed his career and his victory in the American Revolution, at least in part, to fish. Many early presidents, including Thomas Jefferson, Martin Van Buren, and John Quincy Adams, had at least some experience with fishing as boys or men, though sometimes they were frustrating times. "We caught only a few small fish and had the pleasure of rowing a clumsy boat all over the pond," wrote a grumpy nineteen-year-old John Quincy Adams in 1787. It was only natural for the presidents of the 1800s to know their way around a fishing pole, since they came of age in a time when most of America was the frontier, or very close to it.

Abraham Lincoln was once asked what he remembered of the War of 1812. He could only recall one distant boyhood memory: "I had been fishing one day and caught a little fish, which I was taking home. I met a soldier on the road, and having always been told at home that we must be good to soldiers, I gave him my fish." It is one of the very few times we know Lincoln ever spoke about fish. He largely forgot about fishing until he had his own family, and then he took his children to fish on the Sangamon River near Springfield, Illinois.

While fishing in McKean County, Pennsylvania, a couple of years after leaving the White House, Ulysses S. Grant was upset to learn he was fishing out of season.

President Chester Arthur on fishing trip. (National Archives)

He immediately surrendered himself to the closest justice of the peace, who hesitated at passing sentence on the great man. Rejecting any thought of leniency, Grant manfully insisted on paying the full fine for his violation.

The first truly hard-core fisherman in the White House was a mutton-chopped machine politician from New York City named Chester A. Arthur, who served from 1881 to 1885. He was raised a city slicker, but a country boy lurked right below the surface—he helped form a fishing club in Canada that he escaped to as often as he could. "There is nothing I loved more than fishing for salmon," he said.

Arthur was a devoted fly fisherman for most of his adult life. One friend said of Arthur, "No man can pitch a tent more quickly, adorn a camp more tastefully, cast a fly more deftly, fight a salmon more artistically or bring him to gaff more gracefully." For a time he held an Atlantic salmon record of 50 pounds on the Cascapedia River in Quebec. In 1883 he went on a trout-fishing spree during a six-week vacation expedition to Yellowstone. Newspaper cartoonists made fun of his fishing, but as Arthur once said, "I may be president of the United States, but my private life is nobody's damned business."

The next president, Grover Cleveland, was a big-time, 250-pound, heavyweight angler who took his fly rods along on his honeymoon in 1886 after his White House wedding ceremony. His friend Richard Guilder wrote: "Grover Cleveland will fish when it shines and fish when it rains; I have seen him pull bass up in a lively thunderstorm, and refused to be driven from a Cape Cod pond by the worst hailstorm I ever witnessed or suffered. He will fish through hunger and heat, lightning and tempest." Cleveland enjoyed hunting for deer, duck, and quail, but his first love was both fresh and saltwater fishing.

Cleveland even wrote a book about his outdoor passions, called *Fishing and Shooting Sketches.* "There can be no doubt that certain men are endowed with a sort of inherent and spontaneous instinct which leads them to hunting and fishing indulgence as the most alluring and satisfying of all recreations," he wrote, in the flowery prose typical of his era. "I believe it may be safely said that the true hunter or fisherman is born, not made. I believe, too, that those who thus by instinct and birthright belong to the sporting fraternity and are actuated by a genuine sporting spirit, are neither cruel, nor greedy and wasteful of the game and fish they pursue; and I am convinced that there can be no better conservator of the sensible and provident protection of game and fish than those who are enthusiastic in their pursuit, but who, at the same time, are regulated and restrained by the sort of chivalric fairness and generosity, felt and recognized by every true sportsman." In other words, fishermen and hunters are the ultimate conservationists. It was true then, and it's true today, too.

Cleveland seems to have thought about fish as much as politics, and some people thought he was downright obsessed by the creatures. A careful student of both fish and angler, he pinpointed a key flaw in how the fisherman's brain is wired, which could lead to heartbreak: "In many cases the encounter with a large fish causes such excitement, and such distraction or per-

Chester Arthur's presidential hunting and fishing party crossing the Snake River in Yellowstone in 1883. (Library of Congress)

version of judgment, as leads the fisherman to do the wrong thing at the critical instant—thus contributing to an escape."

In 1901, when Cleveland felt fishermen were being slandered for being lazy and foul-mouthed, and dishonest storytellers, too, he sprang to their defense, arguing: "What sense is there in the charge of laziness sometimes made against true fishermen? Laziness has no place in the constitution of a man who starts at sunrise and tramps all day with only a sandwich to eat, floundering through bushes and briers and stumbling

President Arthur's party at Yellowstone. Arthur is seated in the middle. (Library of Congress)

over rocks or wading streams in pursuit of elusive trout. Neither can a fisherman who, with rod in hand, sits in a boat or on a bank all day be called lazy—provided he attends to his fishing and is physically and mentally alert at his occupation."

As to the charge that fishermen told unbelievable fish stories, Cleveland argued, "It must, of course, be admitted that large stories of fishing adventure are sometimes told by fishermen—and why should this not be so? There is no sphere of human activity so full of strange and wonderful incidents as theirs. If non-

Avid fisherman Grover Cleveland lands the presidency in the hotly contested 1884 election as imagined by this political cartoon.

fishers can't assimilate the recital of these wonders, it is because their believing apparatus has not been properly regulated and stimulated." Trust the fisherman, he wrote: "Beyond these presumptions, we have the deliberate and simple story of the fisherman himself, giving with the utmost sincerity all the details of his misfortune, and indicating the length of the fish he has lost, or giving in pounds his exact weight. Now why should this statement be discredited?"

Fishing, Cleveland thought, brings us closer to God: "The real worth and genuineness of the human heart are measured by its readiness to submit to the influences of Nature, and to appreciate the goodness of the Supreme Power. In this domain those who fish are led to a quiet but distinct recognition of a power greater than man's, and a goodness far above human standards. Amid such surroundings, no true fisherman can fail to receive impressions which so elevate the soul and soften the heart as to make him a better man."

Benjamin Harrison, who served a fairly useless single term as president sandwiched in between Cleveland's two, was a lifelong, tobacco-chewing, duck and deer hunter and fisherman who liked to sneak out of the White House for occasional fishing trips to the Adirondack mountains. He spit on his worms for good luck.

Which brings us to the strange case of the President Who Hated Fishing. Until he met one special fish.

On a spring day a hundred years ago, an overweight fifty-seven-year-old man wearing thick eyeglasses and a mustache stood on the bow of a boat off southwestern Florida, gazing into the Gulf of Mexico.

He wore a floppy hat, and held an eight-foot-long wooden harpoon in his right hand, tipped by a steel lance and an iron shank. He scanned the horizon, looking for his target.

The man was Theodore Roosevelt, the former President of the United States. He left the presidency eight years earlier, lost an election bid to return to the White House in 1912, and nearly got himself killed during an epic ill-fated journey down the uncharted, piranha-infested River of Doubt in the Amazon rain forest in 1914. It was such a horrible, disease-plagued trip that Roosevelt estimated it cut his life span short by ten years.

Today, he had one thing on his mind. He wanted to kill a giant manta ray, an ocean creature with a wingspan of up to 29.5 feet and which weighed as much as 4,000 pounds. He had wanted to do this ever since he was a child, explaining, "Killing devilfish [manta rays] with the harpoon and lance had always appealed to me as a fascinating sport." But until now, he'd never gotten the chance.

For most of his life, Roosevelt "detested fishing," in the words of his son Kermit. He didn't mind eating fish, and he would periodically fish for food when he was in the wilderness, but he found the act of poking around streams and hanging around and waiting for a bite to be incredibly boring. He just didn't like to sit still.

As a boy, Roosevelt did a certain amount of fishing, including an 1871 trip to upstate New York that featured his father reading *The Last of the Mohicans* as a bedtime story to the young Roosevelt after a trout meal beside a campfire.

As a ranch owner in the Dakota Badlands in the 1880s with the Little Missouri River on his doorstep, Roosevelt caught fish for the supper table, with some reluctance, as they seemed to gross him out. He reported, "Sometimes we vary our diet with fish—wall-eyed pike, ugly, slimy catfish, and other uncouth finny things, looking very fit denizens of the mud-choked water; but they are good eating withal, in spite of their uncanny appearance. We usually catch them with set lines, left out overnight in the deeper pool."

On the other hand, Teddy Roosevelt loved to hunt big game with firearms, and he was one of the world's first, and greatest, globe-trotting, high-volume hunters, of every conceivable animal, from Rocky Mountain grizzly bears to African rhinos. On hunting trips, when he couldn't find game to eat, Roosevelt settled for fishing, which he sometimes had to improvise. Of an 1884 hunt in Wyoming's Bighorn Mountains, he recalled, "Around this camp there was very little game; but we got a fine mess of spotted trout by taking a long and toilsome walk up to a little lake lying very near [the] timber line. Our rods and lines were most primitive, consisting of two clumsy dead cedars (the only trees within reach), about six feet of string tied to one and a piece of catgut to the other, with preposterous hooks; yet the trout were so ravenous that we caught them at the rate of about one a minute; and they formed another welcome change in our camp fare." As a fisherman, Roosevelt was described by a Yellowstone camping buddy this way: "The President never fishes unless put to it for meat."

But hunting for manta rays with a harpoon in the Gulf of Mexico was a strange, primeval kind of fishing, one that involved bloody combat with a gigantic creature. That's the kind of sport Roosevelt lived for. The bigger the animal, the bloodier the fight, the happier he was.

Roosevelt's target, the manta ray, was an awesome sight to behold. "The extraordinary shape, huge size, and vast power of the big devilfish, or manta, give him an evil reputation, which is heightened by his black coloring," Roosevelt wrote. "In spite of its size the manta is in no way dangerous to man unless attacked; but when harpooned its furious energy, tenacity of life, and enormous strength render it formidable; for it can easily smash or overturn a boat which is clumsily handled, and if the ropes foul an accident is apt to occur."

On his big manta-hunting day in the Gulf of Mexico, Roosevelt absorbed the scenery as he searched for his first victim: "The breeze was light, the sky was glorious overhead, and as the sun rose higher the white radiance was blinding. The tepid waters teemed with life. The dark shadowlike places on the surface marked where schools of fish swam underneath; and to the trained eyes of the professional fishermen in our boat differences that were to me utterly indistinguishable, differences that I could not see even when pointed out, enabled them to tell the species of the fish beneath. Pompano, the most delicious of all food fishes, skipped like silver flashes through the air. Here and there porpoises rolled by."

Suddenly, a member of the crew pointed straight ahead, and a manta ray was spotted. Roosevelt remembered, "It was half a mile off, swimming rather slowly through the water, so near the surface that now and then its glistening black mass appeared for a moment above. The huge bat like wings flapped steadily; occasionally the point of one was thrust into the air."

Roosevelt gripped his harpoon-lance and braced his body for battle. "We stood on the balls of our feet with our knees flexed, taking the movement of the boat," he wrote. "The harpoon was poised in my right hand, which also held a single loop of the rope. Before making the cast I glanced down to see that the rope was not entangled in my feet and would run overboard freely. I steadied myself by gripping the painter, so that I could exert all my strength when I used the harpoon; for I threw with one hand, although the ordinary practice, and doubtless ordinarily the best practice, is to hurl with both hands.

"We saw four or five of the great sea brutes on the surface ahead and to one side of us," he continued, betraying both fascination and disgust with the fishes' appearance. "The black bulk of the manta was a couple of feet below the surface. It was swimming slowly away as the launch, its bow gently rising and falling, came within striking distance."

"Iron [harpoon] him, colonel!" a crewman shouted.

Roosevelt obliged, and threw his harpoon fully two feet, four inches into the fish, penetrating hide, flesh, and bone, and passing through upper part of the heart. Another crewman harpooned it, too. "With a tremendous flurry and a great gush of dark blood the devilfish plunged below and ahead," dragging the launch for half a mile before the men started towing it in. It was a 13-foot-wide specimen, subdued after an eighteen-minute struggle. Roosevelt and his colleagues cheered with joy.

Then they spotted a bigger fish nearby, about two feet beneath the surface, headed head-on toward them, driven, perhaps, by curiosity—or vengeance. Roosevelt readied his harpoon and drew back his throwing arm. "The grotesque black form flapped slowly, the horns were thrust forward; I struck it in the center of the body almost in the exact spot that I had struck the first, and was drenched by the volume of water cast up by the great wing fins," he wrote.

He struck the wounded creature with a second harpoon. "This time the flurry was tremendous and we were drenched with water. We were in a heavy thirty-foot, five-ton launch; yet the devilfish, passing under us and rising, lifted the stern a foot or more upward, and then, sounding, pulled the bow a couple of feet down; and for some little time it actually hauled the launch backward. Then it came to the surface again and towed us in a long three-quarter circle. We began to haul in on both ropes. At last it was near enough for

Theodore Roosevelt prepares to harpoon his first manta ray.
*(*Scribner's *Magazine, September 1917)*

me to dart the lance. As the wicked spade-head drove into its life the huge fish flapped and splashed with such vigor for a few seconds that I drove the lance into it twice again." When the fish expired, they examined it closely and measured it. The manta was a female, measuring a full 16 feet, 8 inches wide.

After a lifetime of waiting, Teddy Roosevelt had his first bloody battle with a manta ray, and his first giant fish kill. He was jubilant, and soon banged out an article on his adventure for the mass-market *Scribner's* magazine. He died less than two years later, on January 6, 1919, at his home at Sagamore Hill, in Oyster Bay, New York.

As president, Teddy Roosevelt made enormous contributions to wildlife conservation. He created the United States Forest Service (USFS), established 150 national forests, 51 federal bird reserves, 4 national game preserves, and 5 national parks, and protected approximately 230 million acres of public land. "Even if Roosevelt had never fished in his life, all anglers would still owe him an enormous debt for his role in the progressive conservation movement at the turn of the century," wrote fishing historian Paul Schullery. "With expert advisors such as Gifford Pinchot [Roosevelt's chief forester], Roosevelt conducted a far-reaching campaign to protect and better manage all kinds of natural resources, from sea birds to watersheds, from archeological sites to elk." That legacy continues to this day.

Roosevelt's 1917 manta ray–hunting expedition to the Gulf of Mexico, conducted on the exact eve of America's entering World War I, had no real scientific purpose. Roosevelt couldn't find a taxidermist to preserve the carcasses for shipment to a museum. This wasn't fishing for food, either. It was strictly for fun.

But it did give Teddy Roosevelt the greatest fishing day of his life.

"Don't think I'll go fishing," said lifelong nonfisherman Calvin Coolidge on the eve of a vacation. "Fishing is for small boys."

That one careless, offhand remark, delivered in 1923, threatened to lose Coolidge the votes of millions of red-blooded American fishermen in the upcoming 1924 presidential election.

Coolidge's Secret Service chief, Edmund W. Starling, an experienced outdoorsman and fly fisherman from Kentucky, went into damage-control mode. He convinced Coolidge to start fishing, and he taught him how.

"Mama! Mama!" Coolidge exclaimed to his wife after catching his first fish, a trout in South Dakota.

"Look what I've caught!" After a few outings, Coolidge was hooked for life. At first, to the distaste of fly-fishing purists, Coolidge preferred fishing with worms, and Coolidge, ever frugal, once asked a guide, "Shall I use the whole worm?" Another time, after losing a fish, Coolidge muttered, "Damn!" Sheepishly, he then quipped, "Guess I'm a real fisherman now. I cussed."

Eventually, Calvin Coolidge learned fly-fishing, and he went off on so many fishing trips that the press made fun of him. He couldn't have cared less. He took fishing and hunting vacations, ranging from six weeks to three months, every summer in 1926, 1927, and 1928, to the Adirondacks, South Dakota, Wisconsin, and Massachusetts.

Coolidge once said his favorite fishing spot had about 45,000 trout.

"I haven't caught them all yet, but I've intimidated them," he wisecracked.

"Where is the salmon?" President Herbert Hoover, Coolidge's successor, asked his secretary.

It was 1929, and a phalanx of press photographers was waiting outside the Oval Office for a first-catch-of-the-year presentation ceremony to the president from a delegation from the Penobscot Salmon Club of Maine.

I sent it to the [*White House*] *kitchen!* replied the horrified, inexperienced secretary, who raced downstairs to rescue the integrity of the salmon's remains.

But the chef had already decapitated the fish and cut its tail off in preparation for cooking. Working frantically at the secretary's direction, he sewed the head and tail back on with needle and thread, stuffed the carcass with cotton, and had the reconstituted creature rushed up to the Rose Garden ceremony.

"The directors of the fishing club, the fish and I posed before 20 photographers," recalled Herbert Hoover, "and each posed for 'Just one more' six times." Before long, Hoover noticed the patch-up job was coming loose, and he struggled to twist the fish into a more flattering pose. "But the cotton kept oozing out of the fish as was proved by the later photographs," according to Hoover. "The fishing club did not use those later editions."

The Case of the Mutilated White House Fish was a kind of metaphor for the next four dismal years of Hoover's term as president, during which he struggled and failed to successfully respond to the Great Depression, earning the scorn and even hatred of many Americans, fishermen and non-anglers alike.

The grumpy Hoover did, however, have one happy outlet to relieve the endless misery of his presidency—fly-fishing. He was a lifelong angler, perhaps the most

President Herbert Hoover with a day's catch, 1929. (Library of Congress)

skilled and philosophical presidential fisherman ever. He fished all across the United States, and while president, he constructed a fishing getaway for himself in Virginia's Blue Ridge Mountains, called Camp Rapidan. The 164-acre retreat was built by a team of 500 marines, featured a dozen cabins, barracks for 250 men, riding stables, and a 100,000-trout hatchery on the riverbank. Hoover escaped there as much as he could.

Hoover claimed to have figured out why so many presidents enjoyed fishing. "I think I have discovered the reason; it is the silent sport," he wrote. "One of the few opportunities given a President for the refreshment of his soul and the clarification of his thoughts by solitude lies through fishing." In retirement, Hoover even wrote a book about fishing, in which he waxed philosophical on his passion: "It is the chance to wash one's soul with pure air, with the rush of the brook, or with the shimmer of the sun on blue water. It brings meekness and inspiration from the decency of nature, charity toward tackle makers, patience toward fish, a mockery of profits and egos, a quieting of hate, a rejoicing that you do not have to decide a darned thing until next week. And it is discipline in the equality of men—for all men are equal before fish. And the contemplation of the water, the forest, and mountains soothes our troubles, shames our wickedness, and inspires us to esteem our fellowmen—especially other fishermen."

Secret Service chief Edmund Starling, the man who taught Calvin Coolidge how to fish, witnessed Hoover's sad decline in office, which even affected his fishing: "As the years went by and the Depression came, President Hoover grew nervous. His hands would tremble as he worked with his tackle. I have seen him catch a fishhook, in his trousers, his coat, and then his hat. It was odd to see this, for he looked like a man without a nerve in his body."

Years after Hoover died in 1964, an old-time Southerner named Wattie Raines recalled the piscatorial pursuits of the thirty-first president along the Rapidan River.

"Yeah, we heard back then that Hoover liked to fish," he said. "We were hoping he would fall in and drown."

Harry S. Truman preferred power walking and poker playing over fishing.

He fished now and then, and posed for fishing pictures to get some votes, but he usually had bad luck. He thought it was because he didn't like to eat fish.

Truman's wife, Bess, was a better angler. One day, according to a friend, Mrs. Truman quickly caught her limit of five trout, and a photographer asked to take a picture. Harry swooped in. "He grabbed those fish and held them up real fast, and Bess was just left standing there off to the side. We all had to laugh about that one. He was a politician. He knew how to get those votes."

Some people were just plain not born to fish.

Richard Nixon was one of these people.

When the future president won election to one of California's U.S. Senate seats, Nixon took a holiday in Florida. His host, Bebe Rebozo, took him on a fishing trip. It didn't go well. Nixon spent the whole time going through paperwork, and barely said six words to his host. "Don't ever send that s-o-b Nixon down here again," Rebozo fumed to the man who introduced them. "He's a guy who doesn't know how to talk, doesn't drink, doesn't smoke, doesn't chase women, doesn't know how to play golf, doesn't know how to play tennis, he can't even fish." He added, "He couldn't bear to kill anything."

Retired U.S. Army General Dwight D. "Ike" Eisenhower must not have known this when he invited Nixon to go fishing two years later.

As soon as Eisenhower won the Republican nomination for president in Chicago in 1952, he asked young U.S. Senator Richard Nixon, his running mate, to join him on a trip to his favorite fly-fishing spot on earth—the Byers Peak Ranch outside Fraser, Colorado. The old general had been going there for years with his buddies. "The gatherings were always a small group of men," recalled Eisenhower. "In simpler, pre-presidential years, this meant cooking, at the most, for three or four. But once I started traveling with Secret Service men, signal detachments, and staff assistants, our simple fishing expeditions became as elaborate as troop movements."

Upon arrival at the ranch, Nixon changed into fishing clothes, his pants hitched ridiculously high over his waist, and Eisenhower taught his protégé how to fly-cast. Or at least he tried to.

"After hooking a limb the first three times, I caught his shirt on my fourth try," Nixon recalled. "It was a disaster." Soon, reported Nixon, "the lesson ended abruptly." Eisenhower looked sad. He could tell that Nixon was all thumbs when it came to casting a fly, and his heart wasn't in it, either. Nixon wrote, "I could see that he was disappointed because he loved fishing and could not understand why others did not like it as well as he did."

As a boy growing up in Abilene, Kansas, Dwight Eisenhower often trotted a few blocks down the Santa Fe Railroad tracks to fish at Mud Creek. His pole was fashioned from a willow tree shoot, his bait was dug-up worms from the family corn patch, and his hooks were five-cent specials from the general store. He caught sunfish, bullheads, carp, and drum at Mud Creek and also at Smoky Hill River and Lyons Creek, twenty miles south of Abilene.

Eisenhower was a burger-flipping, steak-loving, backyard barbecue man who liked to grill-cook for his friends, but he was a big fish lover, too. One of his favorite meals was broiled fillet of trout, and his wife, Mamie, often whipped up dishes made of canned salmon. As an angler, Eisenhower was a fly-fishing purist—"I don't use worms. I want fishing to be a challenge."

At his press conference of October 15th, 1958, Ike explained the appeal of his outdoor hobbies: "There are three [sports] that I like for all the same reasons—golf, fishing and shooting—because they take you into the fields . . . they induce you to take at any one time two or three hours, when you are thinking of the bird, the ball or the wily trout. Now, to my mind, it is a very healthy and beneficial thing, and I do it whenever I get a chance."

During his eight years in office, President Eisenhower went fly-fishing no less than forty times, to Georgia, Colorado, Maine, Vermont, Rhode Island, Maryland, Pennsylvania, and even Argentina. In between grappling with a war in the Mideast, a reelection campaign, tensions with the Soviet Union, and the blossoming campaign for civil rights, Eisenhower's escapes were fishing, golf, and bird hunting. He also enjoyed angling in the trout-stocked streams near the presidential retreat of Camp David. His favorite time to fish was between 6 and 8 P.M., and he carried a small arsenal of flies, including Ginger Quills, Red Variants, and Rio Grande Kings.

Well-wishers sent Eisenhower thousands of gifts of fishing gear to him at the White House. One day, he got a package of hand-made fishing flies from a young prisoner at the State Penitentiary at Rawlings, Wyoming, sent to the president via the governor of Wyoming. The prisoner, said to be a handsome lad and a model inmate, was serving time for manslaughter. Officials were considering a commutation of his sentence.

Eisenhower thought the flies were so beautiful that he packed them right up for his next fishing trip and sent a note to the Governor of Wyoming. "Did the young man get the pardon?" he asked. Yes, replied the governor. His sentence was reduced and he was soon released on parole.

In August 1954, two of the great Republican presidential fishermen had a summit meeting when Eisenhower invited former President Herbert Hoover to come out to Colorado and go fishing with him. "There is a small stream on which we catch ten and twelve inchers, and of course there is always the chance for the occasional big fellow of something on the order of sixteen or seventeen inches," Eisenhower explained in his invitation letter to Hoover. "I assure you that you don't need to be especially terrified at the prospect of living on my cooking for a couple of days. My culinary reputation is pretty good—but my repertoire is limited. It is only after about four days that my guests begin to look a little pained when they come to the dinner table. It is a grand place to loaf and we will have absolutely no one with us except my great friend who owns the place, and possibly my brother Milton. The little stream has many pools that can be fished easily from the bank. Even if you should be compelled to cross the stream occasionally, you will find it remarkably easy to wade. I cannot tell you how delighted I am at the prospect of the two of us having a period together in such a quiet retreat."

Despite the perfect weather and the gorgeous Rocky Mountain setting, the trip did not go well. The grumpy Hoover, who often looked like he'd spent all morning sucking on an extra-sour pickle, thought Ike was both an inept fisherman and a bad cook, who took way too long to whip up breakfast and dinner. He also thought his host was too liberal: he should have been dismantling Franklin Roosevelt's New Deal instead of continuing it, as Ike was doing.

The two presidents hardly caught any fish, and they never went fishing together again.

Eisenhower and his successor, John F. Kennedy, didn't much care for each other—out of earshot, Eisenhower

Left: JFK ocean fishing. (JFK Library)
Right: JFK and his wife, Jacqueline, show off the sailfish he caught on their honeymoon at Acapulco, 1953. (JFK Library)

called Kennedy "Little Boy Blue," and JFK called his predecessor "Old Brass Balls."

JFK did, however, share Ike's love of eating fish, and acquired his passion for seafood as a cod- and clam-chowder–loving child of America's old fishing capital of New England, where his Roman Catholic and family traditions called for eating fish every Friday. Kennedy gave up sportfishing after an epic four-hour fight with a sailfish almost ruined his Acapulco honeymoon, but he got his revenge by mounting the beast for posterity over a fireplace in the West Wing. His successor, Lyndon B. Johnson, did little if any fishing, other than for votes.

As president, Richard Nixon sometimes liked to go fishing with his pal Bob Abplanalp on his buddy's fifty-five-foot yacht, off Grand Cay in the Bahamas. But his passion for angling hadn't grown much from the time of his lesson with Ike—Nixon just sat in a swivel chair with a pole, while aides baited the hook and threw it out in the water. Nixon held the pole and if there was a tug, a crew member reeled in the fish and dropped it in a bucket. A Secret Service man said, "Nixon wouldn't do anything but watch." He was a man who liked to have others do his dirty work.

Gerald Ford, who came into office after Nixon's resignation, had an equally dismal fishing career. "As a youngster I reluctantly fished because my Dad was an ardent fly fisherman for Michigan's trout and he always took me along," Ford recalled to historian Bill Mares in 1994. "When I became old enough to play competitive team sports I lost all interest in fishing. Much to Dad's disappointment. On rare occasions I've done some ocean fishing, but without success. Also, as a good father I took our young children fishing off the piers of Lake Michigan. They inherited my lack of enthusiasm." No wonder this guy didn't win reelection.

When Jimmy Carter beat Ford in the 1976 election, everything changed. A hard-core fisherman was back in charge behind the Oval Office desk.

A little boy stood on a Georgia riverbank.

"What's wrong?" asked his father.

"I've lost the fish, Daddy," said the young Jimmy Carter, who, while struggling with a copperhead bream, had just fumbled a sack containing the day's catch into the current.

"All of them?" asked his father. "Mine, too?"

"Yes, sir."

The boy looked up at the stern, towering figure of his dad, James Earl Carter Sr.

Years later, the younger Carter recalled, "I began to cry and the tears and water ran down my face together each time I came up for breath. Daddy was rarely patient with foolishness or mistakes."

After a long silence, the father said, "Let them go."

The man reached down and put his arms around the boy. The younger Carter remembered, "It seems foolish now, but at that time it was a great tragedy for me. We stood there for awhile."

"There are a lot more fish in the river," the older man said. "We'll get them tomorrow."

"He knew how I felt and was especially nice to me for the next couple of days," said the future President Carter. "I worshipped him."

Casual, backwoods fishing came naturally to America's thirty-ninth president. "I had a fishing pole in my hands as early as I can remember, and would go hunting with Daddy long before I could have anything to shoot other than a BB gun," Carter wrote in his memoir of growing up as a country boy in rural Georgia. He remembered the joy he felt of getting his first baitcasting gear, a hard-earned four-foot-long Shakespeare rod and a Pflueger reel, and a tackle box full of Jitterbugs, Hawaiian Wigglers, and other lures. As governor of Georgia, he first learned how to fly-fish from experts, and his fascination with the sport deepened when he became chief executive of the United States.

Carter's presidency, which often resembled a slow-motion train wreck, was racked by economic turmoil and international crises, and Carter periodically sought relief from his political torments by fishing in the soothing, healing waters of rivers and streams across America. A deeply religious man, Carter may have felt the spiritual power of fishing, too. Ethically speaking, he was a strict catch-and-release man. Discovering fly-fishing, Carter declared, was "one of the most gratifying developments of my life." As a bonus, he often got to enjoy the sport with his wife, Rosalynn, who liked it as much as he did, and she got very good at it. Sometimes, after helicoptering to Camp David and waiting for the press to leave, Carter and his wife would secretly reboard the craft and take a forty-minute flight to Spruce Creek in Pennsylvania for some fly-fishing and peace and quiet.

As a former engineer, the difficulty and technical challenges of fly-fishing appealed to Carter, too. "It levels people out," he once explained to a visitor to his presidential library in Atlanta. "The trout don't give a darn if you're President of the United States or a local farmer or a high-school kid. It's a disciplinary thing, too," he also said, observing the constant need to learn "a little more about how currents behaved, what kind of trout will take a certain kind of fly at a certain time of year or what the temperature of the water does or how a cloudy day affects it or how I can approach a pool without my shadow or silhouette being outlined against the sky. If you read a whole book on it and only come up with one suggestion, it adds to your repertoire and so you become increasingly effective. But then you realize that nobody can ever master the sport."

On another occasion, Carter elaborated, "I really enjoy fishing with a plastic worm for bass in the warm-water areas in South Georgia and Florida. I don't want to derogate that, but at the same time, fly-fishing to me opened up just a new panorama of challenge. Because you had to learn the intricacies of streams, of currents, of water temperature, of different kinds of fly hatches, how to tie your own flies, which you wouldn't ordinarily do in other kinds of fishing, and then try to match whatever fly is hatching off and experiment. It's a matter of kind of stalking, a great element of patience, because the consummate fly-fishers really spend a lot of time observing a pool or a stream or current run before they ever put a fly in the water."

Carter even saw fishing as more rewarding than being the nation's chief executive. "As President, I was able to save with the stroke of the pen a hundred million acres of wilderness area in Alaska," he explained, referring to the Alaska National Interest Lands Conservation Act, which he signed on December 2, 1980, shortly before leaving office, which gave federal protection to more than 157 million acres in the state. "This is the kind of thing that is gratifying to a President, but to be on a solitary stream with good friends, with a fly rod in your hand, and to have a successful or even an unsuccessful day—they're all successful—is an even greater delight."

In 1979, Carter was fishing in a canoe on his farm pond in Plains, Georgia, when a scraggly-looking swamp rabbit swam directly toward him. He smacked the water with his paddle to shoo the varmint sway, a scene that was captured by a White House photographer. The photo, showing Carter defending himself against what came to be known as "The Killer Rabbit," made Carter look ridiculous at a time he was grappling with the Iran hostage crisis, high gas prices, and out-of-control inflation.

At Camp David, the site of his historic 1978 peace negotiations with the leaders of Egypt and Israel, Carter loved to tie his own flies. Just before the peace negotiations began, Carter went fly-fishing in Idaho's Salmon River, where he caught fifty-two trout in a single day. "I took along with me the briefing books that I would be using to prepare myself for the Mideast negotiations, so it was a combination of delightful surroundings, good companions, and also just clearing my mind when I was ready to embark on one of the major challenges of my presidency." The outcome of the grueling eleven-day negotiations was hailed as Carter's greatest success.

In his single term in office as president, Jimmy Carter took no fewer than fifty-two fishing trips, most often at Camp David but also to Georgia, Idaho, Wyoming, Pennsylvania, Missouri, Arkansas, and to Herbert Hoover's old fishing spot on the Rapidan River in

President Jimmy Carter fishing in Alaska, 1980.
(Carter Library)

Virginia. He even squeezed in some bluewater fishing during a 1979 summit with European leaders on the French Caribbean island of Guadeloupe.

The day after Carter was creamed in the 1980 election by nonfisherman Ronald Reagan, he went straight to his workshop and built a varnished wood frame to use as a drying reel for his fly lines.

A few weeks later, as a lame-duck president enjoying his last few weeks as commander in chief, Carter decided that he had to become a better fisherman. So he invited some of America's best fly fishermen to confer with him at Camp David, where they spent the day holding fly-fishing seminars and watching fishing movies.

On the menu that day—trout, naturally.

America's most passionate recent angler in chief was the former World War II fighter pilot, congressman, CIA chief, Republican Party director and vice president George H.W. Bush.

"Bush Senior" enjoyed fishing as a highly rewarding escape from the hectic pressures of being president during the climax of the Cold War, a nuclear coup in Moscow, and the collapses of both the Berlin Wall and the Soviet Union.

"I started fishing at age five," recalled Bush, whose first memories were of fishing off the rocks for tinker mackerel and fishing off his grandfather's boat near the family's home at Walker Point in Kennebunkport, Maine.

Bush remembered "catching mackerel on a lead jig with piece of white cloth for bait," and "trolling with that old green cotton line." He added, "I also fished off the Maine rocks for pollock and for cunner, occasionally small flounder, and I loved jigging for smelt." He became a lifelong fisherman who fished all around North America but always returned to Kennebunkport, his favorite spot, where he spent nearly every summer of his life.

As soon as Bush was elected president on November 8, 1989, he headed off on two fishing trips and one hunting trip, and he spent the last weekend before his inauguration in the Florida Keys, chasing what he called "the wily bonefish."

Bush made no claim to being a technically skilled fly fisherman; he was more of a high-speed, frenetic, full-contact fisherman-for-fun. "He was a troller, a spinner, a bait-caster, racing to ocean fishing grounds in his cigarette boat, *Fidelity*," wrote historian Catherine Schmitt, referring to his twin-engine, blue-on-white racer, specially adapted for fishing and secure presidential communications gear. "He caught bonefish, tarpon, sailfish, mackerel, barracuda, stripers, but rarely ate them. Bush didn't much care for seafood, much less where it came from. He usually threw back his catch or gave it to Secret Service agents." Bush once described the joy of catching-and-releasing a 13-pound bonefish: "You take them in, weigh them, and hold them in the water and stroke them so that they can escape without some shark getting them. It really is wonderful."

Bush explained why he loved fishing so much, and like other presidents, his description sounded almost like a form of prayer or meditation: "It totally, totally clears your mind. You relax. It's not just catching the fish; it's the background, the environment, the beauty of it all. You can get mesmerized by those waves and this clear surf. So I get a kick of it just catching or trying to catch a fish, but from being in this setting." On another occasion, he noted, "I love the water. I can concentrate and forget all my worries. I count my blessings while fishing." He summed up his feelings by explaining, "I am hard-pressed to think of an outdoor activity that teaches you more about the beauty of nature and the joy of family and friends than does fishing." He once compared fishing to being in "heaven." Out on his boat, he said, "You're in another world." In the heat of his summer 1988 campaign for president, while out on his boat fishing off Kennebunkport, Bush delivered the unforgettable tough-guy line, "Those bluefish are dead meat."

Father and son fishing. (National Archives)

Bush's popularity among anglers was of no help to him in his 1992 presidential reelection campaign, which he lost to longtime nonfisherman Bill Clinton.

In 1997, as an ex-president, Bush took a fishing trip to Canada's Northwest Territories. He enjoyed it so much that he sent a long, heartfelt message of thanks to the local newspaper. "Way above the tree line, the fast-flowing Tree River pours its rushing green-gray waters into the Arctic Ocean, about a mile or two from where I fished for char," Bush wrote. "If thirsty, you can cup your hands and drink of these pristine waters.

Yes, there are some mosquitoes around, but not enough to detract from the joys of fishing. Even a mild breeze seems to keep the critters away. This year the weather was perfect. We fished in T-shirts, needing a sweater or jacket only in the early morning or late afternoon. I had 43 fish on my fly rod, only to bring two into the shore. Don't laugh; I was proud to have kept the fly in the water, having the thrill of that many fish, even for a moment, on my No. 9 rod. I am a very happy and a very lucky man now. Because of time spent fishing and the chance that fishing gives me to relax and think freely, now more than ever I see clearly just how blessed I really am. I served my country. I have a close family and a wonderful wife, and yes, I went to the Tree River and caught char. Tight lines to all you fisherman! Submitted by this most enthusiastic amateur, to whom Canada has given such joy."

Ten years later, Bush's son, President George W. Bush, asked his dad for a favor, and he was happy to oblige.

"Just let me know what you need, son," said the elder Bush.

The president wanted his father to host President Vladimir Putin of Russia at the Bush family home in Kennebunkport, and take him out fishing. The father readily agreed.

The younger Bush remembered, "When Putin arrived on July 1, 2007, Dad met his plane at the airport in New Hampshire and accompanied him on the helicopter ride to Walker's Point. Then he took both of us for a speedboat ride. Although initially startled by the idea of an eighty-three-year-old former President driving the boat at top speed, Putin loved the ride. (His interpreter looked like he was about to fly out the back of the boat.) The next morning, we had a long conversation about missile defense, in which we found some common ground. We then went fishing. Fittingly, Putin was the only one who caught anything."

As president, Barack Obama did not go to Camp David that much, and he spent little time fishing there. But he did want to learn how to fly-fish, so when his schedule took him to the Big Sky Country of Montana in August 2009, he penciled in a two-hour afternoon break to take a fly-fishing lesson on the beautiful East Gallatin River.

"The weather on the day of fishing with President Obama was horrendous," said Dan Vermillion of the Sweetwater Fly Shop, Obama's guide and fishing teacher. "The fishing conditions were very difficult, with gusty winds and a steady downpour. It was the type of weather that would have sent most of us to the

President Barack Obama enjoys fly-fishing in Montana, 2009. (Pete Souza, White House photo)

bar." But, Vermillion reported, Obama "persevered through the weather." He described Obama as "a very quick learner," who "was really interested in learning about fly-fishing" and had practiced some casting moves at Camp David before the trip. Vermillion reported, "He ended up with several strikes and fish that were hooked. However, we did not land a fish. If time allows, I think he will become a very good fisherman. He is obviously a very good athlete."

"The thing that impressed me most about President Obama was his sincerity," Vermillion said. "He did not just want to catch fish. He really wanted to understand why we were using the flies we were using, why we were fishing the water we were fishing, and what life cycles of trout in the East Gallatin were. I think he will catch a lot of fish in his day."

There's one thing that all the fisherman presidents learned from hard experience—there are some critics that you just can't satisfy.

In Jimmy Carter's case, it was his own mother.

"You may be president," she told him, "but you still haven't learned how to catch fish."

9
The Power of Fishing

Man's life is but vain; for 'tis subject to pain,
And sorrow, and short as a bubble;
'Tis a hodge-podge of business, and money, and care,
And care, and money and trouble.
But we'll take no care when the weather proves fair,
Nor will we vex now though it rain;
We'll banish all sorrow, and sing till tomorrow,
And angle and angle again.

—Izaak Walton, *The Compleat Angler*

In the night I dreamed of trout-fishing; and, when at length I awoke, it seemed a fable that this painted fish swam there so near my couch, and rose to our hooks the last evening, and I doubted if I had not dreamed it all. So I arose before dawn to test its truth, while my companions were still

sleeping. There stood Ktaadn [Mount Katahdin, highest mountain in Maine] with distinct and cloudless outline in the moonlight; and the rippling of the rapids was the only sound to break the stillness. Standing on the shore, I once more cast my line into the stream, and I found the dream to be real and the fable true. The speckled trout and silvery roach, like flying fish, sped swiftly through the moonlight air, describing bright arcs on the dark side of Ktaadn, until moonlight, now fading into daylight, brought satiety to my mind, and the minds of my companions, who had joined me.

—Henry David Thoreau, "Ktaadn"

It is a great time to be an American Fisherman.

We have an amazing history. We helped build this nation. Today we are a strong, growing, and interconnected community.

Depending on who's counting, there are 33 million to 46 million anglers in America today, and nearly one-third of us are women. According to the latest U.S. Fish & Wildlife Service survey, more than 47 percent of first-time fishing participants are female. More than 2 million new fishermen join the ranks every year and have their first fishing experience, and lots of them are children. Every year brings new advances in fishing technology—and fishing conservation. The $36 billion spent on the sport each year provides lots of jobs and income for Americans. It's never been easier or more fun to learn how to fish, or how to be a better fisherman. The Web and cable TV are chock-full of great fishing shows and tips.

The vast majority of us are responsible, ethical fishermen, who practice catch-and-release fishing, or fish for food that we eat ourselves or share with others. The taxes, licenses, and fees we pay help keep America's waters clean and our forests pure. Fishermen, along with hunters, are among America's strongest conservationists. We don't run around bragging about it much, but we're proud to do it, because it's the right thing to do.

Despite continuing hazards, commercial fishing is getting safer, thanks to the hard work of many professional fishermen. Although America has experienced severe episodes of overfishing and the near death of several important species over the years, many stocks have rebounded through careful stewardship, and we are on track to end overfishing for good.

The benefits of fishing are incredible. Fishing is one of the most easily accessible outdoor sports. Practically anyone, of any age, income level, or fitness ability, can easily become a fisherman. You can fish all by yourself or with a bunch of friends and family. Fishing can help keep you physically fit, by getting you into the outdoors, fresh air, and sunshine. I think it makes us happier and helps us age gracefully. Fish are super-nutritious, and a diet rich in fish can protect your heart and help you live longer. Fishing can help reduce stress. It is a perfect escape from our overdigitized, pressurized world. Fishing is like prayer or meditation, and it can bring us closer to God and nature.

In fact, fishing has the power to heal injured bodies, renew souls, and even to save lives.

One day in 2004, a retired Navy captain from Port Tobacco, Virginia, named Ed Nicholson was recovering from an operation at Walter Reed Army Medical Center in Bethesda, Maryland. He spent thirty years in the Navy as a Surface Warfare Officer, then worked for a decade as a defense contractor. He was a prostate cancer survivor, an outdoorsman, and an avid fly fisherman.

He noticed a soldier next to him who had lost an arm in Iraq. The soldier was enduring a tough daily regimen of physical therapy and medications. Then he saw a young man in his early twenties, another veteran and patient, with no legs, sitting in a wheelchair, with a baby in his arms and his wife pushing him down the hallway. He witnessed scores of other disabled and wounded veterans, including a number of amputees, grapple with their injuries. They were pretty much stuck in the hospital, and trapped in the routine of being cared for as survivors.

Nicholson remembered, "Other than being in Vietnam and seeing people in the process of getting hurt, I never really had a full appreciation for the recovery part and what happened after they came home. My recovery was nothing compared to what they were facing. It planted the seed that maybe there's something I could do."

Then Ed Nicholson thought about fly-fishing. What if we could get these guys outside, he thought, into the fresh air and great outdoors, even just on the lawn of the hospital for starters, and we taught them how to cast? What if we taught them how to tie their own flies, and taught them how to get out in a stream and try catch-and-release fishing for the very first time? Maybe, he thought, it could cheer them up, give them a welcome diversion and a new hobby, and help them strengthen things like their fine-motor skills, concentration, eye-hand coordination, balance, range of motion, camaraderie, and sense of well-being and achievement. "I should take a couple of these guys fishing with me," he realized. He talked to the military brass at the hospital, who thought it might be a good idea.

Nicolson asked permission to take a soldier out fly-fishing one weekend. Permission granted. The soldier was overjoyed—and transformed. The effect was so powerful that Nicholson was asked if he could take seven more vets out fishing the next weekend. One thing led to another, and today, Ed Nicholson is founder and president of a thriving national nonprofit outfit called "Project Healing Waters Fly Fishing." It has 140 chapters in 46 states and internationally, many of them plugged into veterans hospitals and fishing groups like Trout Unlimited. Project Healing Waters has changed the lives of thousands of veterans—through fishing.

"The goal is simple," Nicholson explained. "You have a guy who lost a leg, he's in physical therapy. We get him out there wading a stream, he gets a boost. Or a guy who lost an arm, we start him casting. He has a chance to use his new arm and actually do something that's enjoyable. It's good physical therapy, and then there's the emotional part." The Project Healing Waters program started as a basic day trip, but now includes classes on fly-fishing, casting, tying, and rod building along with clinics for both beginners and for experienced fishermen facing new physical challenges.

The program is available to any disabled veteran who wants to participate. All expenses are paid.

The serene, relaxing, rhythmic, and focused experience of fly-fishing, it turns out, is intensely therapeutic. "When we started this, I thought it would just be great to take some of these guys fishing," Nicholson said. "But I can't tell you the number of times someone has come up to me and said: 'Thank you. This program saved my life.'" He added, "For many of these guys, fly-fishing has become a tool of recovery, both mentally and physically." "We don't just take people fishing. We build relationships, and within that comes the healing. It transcends fly-fishing . . . the friendships are where people heal," he said. "Between the pain, the medication, the realization that their life has been changed, they're doing something that gives them a great deal of pleasure and that they can look forward to."

Many veterans report that the program has changed their lives for the better and even saved their lives. Ed Nicholson gets appreciation not only from veterans but from spouses, who thank him for giving them their loved ones back.

For example, Jan Bradbury, the wife of participant Chuck Bradbury, said, "It's a wonderful thing to behold. Chuck is a Vietnam combat veteran, a former door gunner on a Huey, and was wound a little tight, and withdrawn. As we quietly celebrate his homecoming exactly forty-five years ago today, and reflect on all that has meant to him over the years, and how all that has affected him over the years, I can honestly say, with all conviction, that Project Healing Waters has effected more positive, lasting changes in my husband than any other program, ever. He is so much calmer that now it's possible for his psych meds for PTSD to be reduced. And it's not just the effect of the meds, it's a calmness that is pervasive, through and through. Calmness that goes from head, to hands, to heart. He rests well, he is nowhere nearly as restless, and as a result there is more calm, a restfulness in our home, our marriage, our life together."

Valerie Takesue was a U.S. Army major who was sent from a hospital in Korea to the Warrior Transition Unit at Fort Meade, Maryland, and medically retired as a major after twenty-six years of service. Suddenly, she had to cope with the sudden end of her military career as well as physical challenges. "When I came back from Korea, I didn't want to do anything," she said. But of Project Healing Waters, she said, "It has given me a positive focus in life after the military. It got me out doing things besides being upset all the time. Fly-fishing actually gave me a focus to concentrate on something else, a lot of the negatives have changed to positives." She added, "It's been very therapeutic. I've been so limited in what I can do; fishing is not too hard. It's been wonderful for my psyche; it's been wonderful for me physically, too."

Kyle Chanitz, an Army veteran who served three tours in Afghanistan, sustained brain injuries from two concussions. After his 2011 discharge, he had nine seizures and found his thoughts dominated by Afghanistan. The twenty-nine-year-old New Yorker moved to Roanoke, Virginia, and entered a post-traumatic stress disorder program, where he attended a Project Healing Waters session. He was fascinated, and that night he ordered a fishing kit. "When you're fly-fishing or fly tying you don't think about anything else but fly-fishing," he reported. "It's nice to get your mind off everything." He found the sensation of catching a fish with his own fly to be indescribable. On an overnight trip to Back Creek, near Hot Springs, Virginia, he caught fourteen trout and five bluegills. "This is probably the happiest I've been since the Army," he said.

On the South Holston River in Tennessee, U.S. Army veteran Charlie Pannell, who lost a leg and sustained many other injuries from an improvised explosive device in Iraq in 2008, recalled his first experiences fishing for trout. "All you could hear was the sound of the river. I remember thinking, this is a new passion you can come to love." Former Marine Nate Moore, who was wounded in Afghanistan in 2010, caught his first trout on a fly rod, and noted, "This program does a lot more than just help the servicemen; it helps the entire family heal."

Russ Marek of Viera, Florida, a former staff sergeant in the Army, lost his right arm and right leg and endured a brain injury and burns over 20 percent of his body when a roadside bomb detonated beneath his vehicle on September 16, 2005, in Iraq. He learned the basics of fly-fishing from Jason Redler, a Healing Waters volunteer and Gulf War veteran who copes with post-traumatic stress disorder. "It helps me out, and it helps someone else," said Redler about the relationship. "It helps both of us out." Marek, his student, said, "It's a new challenge. It expands your imagination. I feel comfortable. I feel happy that they are teaching us something new." Marek, who has a prosthetic leg and a prosthetic arm, uses a fly-tying device called an Evergreen arm, which has magnets that help fasten the tiny, complex pieces into position. "You've got to be very imaginative for these things," he explained. "It takes your mind off everyday struggles. For newer veterans coming home, it will take their minds off war issues."

"This program," said disabled Vietnam-era vet David Hall of Project Healing Waters, "is the best thing that ever happened to me."

There's another great fishing program that is changing people's lives for the better. It's called Casting for Recovery, and it helps breast cancer survivors—through fly-fishing.

It started in 1996 in Manchester, Vermont, when Gwenn Perkins, a former Orvis company casting instructor, and Dr. Benita Walton, a breast reconstruction surgeon, suggested that breast cancer survivors might benefit from the physical and spiritual benefits of fly-fishing—and from being pampered in the beauty of nature. In 1998, they and their friends organized four fishing retreats, and the nonprofit organization has grown ever since, helping thousands of women recover and heal.

Typically, a Casting for Recovery retreat lasts two and a half days in a beautiful fishing destination. Lodging and meals are free. Participants receive medical education, group discussions, and support from professional facilitators. The last half day, devoted to catch-and-return fly-fishing, is the payoff. Women are fully outfitted with fishing gear, given instructions, and set loose with individual guides to savor the joys of nature and catch and release some fish.

For many women, it is an experience of pure joy and peace. “It was almost hypnotic,” marveled breast cancer survivor and Casting for Recovery participant Johanna Thompson, “It was something where everything else melted away. You focus on the rhythm, you say, ‘I focus on me. Look at how healthy I am. I’m surrounded by all this beauty.’” JoAnne Brown, a middle school teacher, reported, “It focuses you in on the moment. You’re not thinking about what’s going to happen tomorrow, what’s going to happen next year. It’s just you and the fish. It’s so cool. There is something about standing in the river that makes it more tactile, more real. You’re just a part of everything. I’m a part of nature.” Retired teacher Mary Roberson, who attended a retreat in the mountains of Wyoming, called the experience of fly-fishing “magical.”

Nancy Shoemaker, another breast cancer survivor, especially enjoyed the process of learning to tie flies. “It was such a Zen-like experience,” she remembered. “You have to pay so much attention, it’s such intricate, small work that you can’t think about other stuff.”

Whitney Milhoan, the executive director of Casting for Recovery, explained that the benefits are physical as well as emotional and spiritual. She noted, “Fly-fishing is a physically nondemanding sport, but the motion of casting a rod can be a good way to just encourage mobility in the upper arm and body for women who might have had surgery or radiation as part of their treatment for breast cancer. The gentle rhythm of fly-casting can help heal the joints and soft-muscle tissue damaged by radiation and/or surgery."

Fly-fishing offers stress relief and calm, and the social interactions give women a chance to bond, reenergize—and laugh together.

"I hadn't fished since sixth grade, when I was in Girl Scouts," said Sally Schultz, a legislative aide and breast cancer survivor who renewed her love of angling on a Casting for Recovery retreat on Nebraska's Snake River. "It was a blast."

Fishing continues to be an important part of my family's life, and I expect it always will. I've introduced all my children to fishing, and I sure hope they'll continue on with it if they so choose, and teach their own kids how to fish when they're young.

Ever since the first Native Americans began fishing thousands of years ago, I think there has always been a mystical connection between American fishermen and the water. It is a spiritual bond we have with each other, and with the bounty of life that God has given

Trout Fishing in Montana. (Robertson Family Collection)

us. I believe that the Native Americans felt that bond, as did George Washington, Theodore Gordon, Cornelia Crosby, Franklin D. Roosevelt, Dwight Eisenhower, Joan Wulff, Ed Nicholson, and all the great Americans who have enjoyed the gift of angling.

The Bible tells us we have a responsibility to safeguard that gift, and to pass it on to the future generations. We have a duty to fish ethically and responsibly. It is a promise we must keep.

As we enjoy the bounty of God's creation, let us always keep in mind the instruction he gave us in

Marcus Luttrell, me, and Adam LaRoche fishing for trout in Montana. (Robertson Family Collection)

Genesis: "Be fruitful, and multiply, and replenish the earth."

THE FISHERMAN'S PRAYER

By Homer Circle

God grant that I may fish

until my dying day;

And when at last I come to rest,
I'll then most humbly pray;
When in His landing net
I lie in final sleep;
That in His mercy I'll be judged
as good enough to keep!
Amen

Acknowledgments

We thank our families; our editor, Peter Hubbard, and his colleagues at HarperCollins including Nick Amphlett, Heidi Richter, and Lauren Janiec; our agent, Mel Berger, and his assistant David Hinds at William Morris Endeavor; our copy editor Tom Pitoniak; Jack Vitek of the International Game Fish Association; Dave Precht and Helen White of Bassmaster Inc.; historian Paul Schullery; and Melanie Locay of the New York Public Library.

Appendix A: Great Moments in American Fishing

The history of American angling is the story of countless great moments—of discovery, excitement, frustration, and joy. Often they are simple moments between friends, between parents and children, or just between an angler and a river. Here are some other great moments in the history of American fishing.

Early 1800s: The modern multiplying, baitcasting reel is originated by Kentucky watchmakers George Snyder and Jonathan Meek, who applied their expertise on intricate gearing to fishing reels.

1811: Julio Buel accidentally drops a spoon into a lake in Vermont from his rowboat. To his surprise, a large bass appears, gobbles the spoon, and vanishes. Intrigued, he experiments with spoon designs and creates the first spoon-style fishing lure; he eventually launches his own factory.

Mid-1800 on: Bamboo rods achieve mass popularity. The innovative designs of master rod maker Paul Young become a gold standard.

Late 1850s: the sport of fly-fishing reaches many spots in the American West.

1865: Dr. C. C. Abbott catches a 4-pound, 3-ounce world-record yellow perch in New Jersey, establishing the longest-standing freshwater record.

Post-1865: Recreational fishing takes off in the United States.

1868: Hiram Leonard pioneers a highly popular, and efficient, split-bamboo fly rod design, and transforms the art of rod building by using industrial manufacturing methods.

1871: Smithsonian Institution director Spencer F. Baird is named the first head of the new U.S. Fish Commission. He oversees a historic "fish shuffle" of hatcheries and species: eastern brook trout were transplanted to the Rockies and Far West; rainbow trout from the West Coast were sent to the Rockies, Midwest, and East; brown trout from Europe were brought to America; East Coast shad and striped bass were brought to California; and largemouth and smallmouth bass were stocked nearly everywhere.

1872: The American Fish Culturists Association appropriates $15,000 for the U.S. government to begin fish culture development, expanding the federal role in aquaculture.

1874: Seth Green transplants the hardy California (rainbow) trout to the eastern United States, hoping it can supplant the more pollution-sensitive brook trout. It works, and rainbow trout thrive across the East and the upper Midwest.

1874: Charles F. Orvis of Vermont manufactures the first modern fly reel, a breakthrough reel and fly design described by historian Jim Brown as the "benchmark of American reel design."

Late 1800s–early 1900s: Freshwater fishing is transformed with lighter-weight gear like bamboo rods;

lines made of linen, cotton, and silk; and the fixed-spool spinning reel. Much longer casts are now possible.

1893: Mary Ellen Orvis Marbury exhibits her own hand-tied flies at the World's Columbian Exposition in Chicago.

1905: Lee Wulff is born; he will become arguably the most influential and popular American fisherman. He pioneers the use of light tackle to take game fish, innovates the fishing vest and the hair-wing dry fly, produces TV shows, books, and articles, lands a 10-pound salmon on a No. 28 hook, and, most important, with his wife, Joan, becomes a leading champion of catch-and-release fishing.

1905: Theodore Roosevelt sets aside 150 million acres of timberland, doubles the number of national parks, and creates fifty game refuges and sixteen national monuments.

1909: Norwegian-American inventor Ole Evinrude introduces the first commercially successful outboard motor, a 46-pound, 1.5-horsepower, two-stroke version.

1912: Former dentist Zane Grey publishes blockbuster Western novel *Riders of the Purple Sage*, becomes the most popular writer in America (60 novels, 13 million copies), and a globe-trotting, record-holding big-game fisherman who spends more than three hundred days each year fishing, usually in the Santa Catalina Channel, off Southern California, on his 52-foot cruiser, Gladiator.

1913: William Boschen becomes the first person on record to land a broadbill swordfish (a 358-pounder) with hook and line, off Catalina Island in 1913. His fishing partner, boat captain George "Tuna George" Farnsworth, works with Boschen to create the first reel with an internal star drag.

1913: Julius vom Hofe of Brooklyn, New York, invents a reel with an internal drag, which enables catches on lighter line and tackle and helps spread big-game fishing to the Atlantic.

1922: Sportsmen form the Izaak Walton League, one of America's oldest conservation groups, which helped launch the first congressionally funded wildlife refuge, on the upper Mississippi River.

1929: Winston Churchill visits the Tuna Club of Avalon, on Catalina Island, reels in a 125-pound marlin, orders a scotch and soda, lights a cigar, and says, "I see why you chaps enjoy this, it's great fun."

1930s: Recreational tuna fishing takes off, with tournaments being held from the Bahamas to Nova Scotia.

1930s–40s: President Franklin D. Roosevelt's New Deal dam-building and river and water management projects trigger a sharp rise in forage and warmwater game fish populations, making fishing more accessible to millions of Americans.

1932: Captain Jay Gould catches what may be the biggest fish ever hooked and landed, a manta ray that measured 19 feet, 9 inches from wing tip to wing tip and weighed an estimated 5,500 pounds. The manta was hooked off Florida on a shark hook attached to 1,200 feet of half-inch rope, and a 20-ton crane was used to hoist it from the water.

1932: The Woman's Flyfishers Club is founded in New York's Catskill Mountains by Julia Fairchild, who served as its president for the next thirty-nine years.

1939: At a meeting at the American Museum of Natural History, the International Game Fish Association is established. Michael Lerner is a key force in founding the group, which grew from discussions he had with Ernest Hemingway and others. The IGFA becomes a leading force for conservation and fishing ethics, and is now the global authority on fishing record-keeping. Lerner's wife, Helen, is an accomplished angler as well, supporting marine conservation and becoming the first woman to take a bluefin tuna on the European continent, the first to take nine tuna in one year, the first to catch a broadbill in both the Atlantic and Pacific oceans, and the first to take four different species of marlin.

World War II: Three women, Beulah Cass, "Bonefish" Bonnie Smith, and Frankee Albright, become influential fishing guides in the Florida Keys, pioneering women's involvement in a field dominated by men.

Post–World War II: Sportfishing is transformed by new technologies and new materials like nylon lines, fixed-spool reels, plastic lures and lines, aluminum, carbon fiber, fiberglass, titanium and graphite rods, transistorized sonar, and GPS devices. Some of this same technology benefits commercial fishing also.

1946: Palm Beach, Florida, boat builder and conservationist John Rybovich Jr. and his brother Tommy establish the prototype of the modern sportfishing boat, the 34-foot Miss Chevy II, featuring an elevated foredeck, flybridge controls and spacious cockpit aluminum outriggers, transom doors, and Rybovich fighting chairs.

1950: The Dingell-Johnson Act authorizes an excise tax on fishing equipment, to fund fisheries management.

1955: Ernest Schwiebert publishes his influential book, *Matching the Hatch,* a guide for matching artificial trout flies to natural insects, while an undergraduate at Princeton University. More books followed, including *Nymphs* (1973) and *Trout* (1978).

1955–60: Physics professor Buck Perry of North Carolina pioneers "spoon-plugging," an innovative system of exploiting the deepwater habits of species like bass and walleye that transforms warm-water angling and marks the dawn of the age of structure fishing.

1957: Carl Lowrance adapts Navy technology to fishing by introducing his sonar depth-sounder/fish-locator, launching the age of the era of electronic fishing.

1959: Trout Unlimited is formed. Today it is a national nonprofit organization with 150,000 members dedicated to conserving, protecting, and restoring North America's coldwater fisheries and their watersheds.

1962: Minnow-shaped plugs, inspired by the designs of Finnish backwoods commercial fisherman Lauri Rapala, are launched in the United States by the Normark Corporation. Rapala's designs mimicked the jerky movements of a weak or sickly minnow, which attracted predatory fish. Eventually, millions were sold, becoming the most popular lures in fishing history.

1964: The Wilderness Act of 1964 becomes the main mechanism for protecting the American wilderness; it now protects more than 100 million acres of federal land, half of it in Alaska.

1968: Insurance man Ray Scott of Montgomery, Alabama, founds the Bass Anglers Sportsman Society (B.A.S.S. Inc.) and launches the era of big-money fishing competitions with the first professional bass tournament, on Beaver Lake, Arkansas. Sleek, fast bass boats become popular. By the end of the 1970s, largemouth and smallmouth black bass edge out both trout and panfish as the favorite fish in the United States as measured by the number of anglers. The Bass Anglers Sportsman Society becomes a $30-million-a-year business with 650,000 members and its own magazine and television show.

1968: The American Museum of Fly Fishing is established in Manchester, Vermont.

1976: The Magnuson-Stevens Act is passed by Congress to govern marine fisheries management in federal waters. The act extends U.S. jurisdiction from 12 to 200 nautical miles from shore; over the next forty years overfishing is reduced and overfished stocks are rebuilt for a number of species.

1976: Sugar Ferris starts Bass'n Gal, the first bass fishing tournaments for women. It grew to 32,000 members in clubs and held nine tournaments a year but ceased operations in 1997. Ferris said, "After having heart surgery, I am just worn out from the never-ending effort of trying to get and keep corporate sponsorship. Sponsors don't want to give a long-term commitment, but they want to tell you how the tournaments are run. My grief is for all the women who loved the sport."

1996: The International Federation of Black Bass Anglers (IFBBA) is established to develop the growing interest that many African-Americans have in fishing.

1996: The five hundredth anniversary of the first book about fishing published in the English language, *A Treatyse on Fisshynge Wyth an Angle,* by Dame Juliana Berners.

2013: The International Game Fish Association honors Heather Michelle Harkavy, at sixteen the youngest female angler ever to achieve one hundred world-record fish catches.

Appendix B: The Evolution of a Fly-fisher

By Joan Salvato Wulff

Joan Salvato Wulff is the first fady and master teacher of American Fly Fishing. In this essay, which she graciously shared with us, she speaks of her evolution as an angler.

Trout season is about to open and trout fishing in the Catskills is always exceptional. Our beautiful clean streams produce the myriad aquatic insects that are trout's primary food. Get out your imitations of mayflies, caddis flies, stone flies, and midges!

Realize that fly-fishing is a lifetime sport. You can delve into its different facets as deeply as you like, to keep up your interest and to enjoy an ongoing evolution.

My own evolution started as an innocent; I had no obvious predatory skills. I saw interesting patterns on the surface water of trout streams, but had no concept of "reading" where trout might choose to lie and feed *under* the surface.

Having come to the sport by way of tournament casting, that skill allowed me to "cover" all of the water, to learn where the fish might lie to fulfill their need for food, comfort, and safety.

Over the ensuing seventy-nine years I've been lucky enough to fish for most of the game fish species of the world that are willing to take a fly, and I have come to realize the evolution I have gone through. The first three stages of the evolution are familiar: 1. How Many? 2. How Big? 3. How Difficult? These three stages play to our competitive nature, measuring ourselves against others whenever there is something to count. And they may stay with us forever in terms of a particular species, especially in the "difficult" category. I'm still after a 15–20 pound Permit, a saltwater species.

It was Lee Wulff who raised my consciousness to Stage Four, which is about more than catching fish. It is about "giving back" to the resource; looking at the sport from the point of view of the fish; preserving their gene pools as well as their habitat. We do this by practicing Lee's mantra of "Catch and Release."

This is when we join the conservation organizations: Trout Unlimited, The Federation of Fly Fishers, and other game fish groups. These organizations give more meaning to the sport in that you are involved with people who share your values—both in the sport and outside of it.

The stages beyond Stage Four are the golden ones. Stage Five is *just being there.*

My favorite place to fish is wherever I am. My favorite fish is whatever I'm fishing for. As my experience has broadened and I've been exposed to the characteristics of different species of game fish, always in beautiful places and in clean water, I have come to love and admire all of them. It's like partaking of good food. Think of how many different foods you really love and appreciate each time you enjoy them. So it is with fly-fishing.

Stage Six: It may be hard to believe but *catching* fish becomes less important. I can end up fishless and still have had a good day if there were fish to be caught and my presentations were good ones. Also, I can now fish "through" a companion and be as happy about their catch as they are, because I know the challenges, and feelings of joy and satisfaction that they are now feeling. I call this stage "maturity."

Stage Seven: Replacing yourself by teaching others (especially grandchildren and other youngsters) the joys and responsibilities of the sport.

I think I'm coming into Stage Eight, through an ardent angler friend's practice. The electric feeling of the strike is the most memorable part of recalling a catch. When the fish are easy to catch, "cut off the curve of the hook." The strike will still be memorable and the fish is free, without the stress of being played, to go on living and take a fly again.

Tight lines!

Appendix C: The Thrill of Northern Fishing

By President George H.W. Bush

In 1997, one of our most enthusiastic American fishermen-presidents, George H. W. Bush, took his family on a fishing trip to Canada's Northwest Territories. He had such a good time that he sent a thank-you note to the people of the town. We're real proud that he shared the note with us, too. Thank you, Mr. President!

I love the Tree River. Way above the tree line, the fast-flowing Tree River pours its rushing green-gray waters into the Arctic Ocean, about a mile or two from where I fished for char.

As the waters race over the boulders and rocks, you can catch an occasional glimpse of the majestic char, struggling to continue their fights against the current, their quests to reach their destiny, upriver quest.

If thirsty, you can cup your hands and drink of these pristine waters.

Yes, there are some mosquitos around, but not enough to detract from the joys of fishing. Even a mild breeze seems to keep the critters away.

This year the weather was perfect. We fished in T-shirts, needing a sweater or a jacket only in the early morning or the late afternoon. The weather up there is variable, and it can get very wet and very cold even in August, but not this year.

There were a lot of char in those fast-running waters, a lot of big, strong fish. My 13-year-old grandson, Jeb, from Miami, Fla., got a 25–30-pound fish on his Magog Smelt fly—a brown, wet fly that was very productive over the course of our whole trip.

He fought the fish for 45 minutes, following our guide, Andy's, instructions to perfection. The big red, finally tiring, came into the shallow waters just above some rapids, and then with one ultimate surge of energy, he flipped over the edge of the pool into the white-water rapids, broke the 20-pound test tippet, and swam to freedom.

My grandson, not an experienced fly-fisherman, had fought the fish to perfection. He did nothing wrong.

All the fishing experts who were watching told him so, but these big fish are strong and tough and they never give up.

I had 43 fish on my fly rod, only to bring two into the shore. Don't laugh; I was proud to have kept the fly in the water, kept on casting, having the thrill of having that many fish, even for a moment, on my No, 9 rod. I used an L.L. Bean reel.

As for the flies, I found that the Mickey Finn, the Blue Charm, and the Magog Smelt all worked well. So did some others, the names of which escape me even as I write.

I tried some dry flies, but they produced zilch in the way of action.

I found that I got most of my fish on when the fly was drifting downstream, though I got two or three hits the instant the fly hit the water. One pool was narrow, right next to the fast part of the water. I'd throw the fly into the white-capped waves, and it would be rushed by the current into the pool. When it left the raging water and hit the more placid pool, the fish would strike.

I did better on getting the fly unhooked from the rocks this year, though I did lose a tiny number of flies when they were claimed by some especially craggy rocks.

I learned that the way to get lots of fish on the line is to keep the hook in the water. Obvious? Well, maybe, but a lot of fishermen seem to hang out waiting for someone else to catch one before they'd do serious fly-fishing.

The rocks were very slick, and, at 73 years of age, my balance is less than perfect. Put it this way: I can't turn very well and I slip a lot. The felt-bottomed boots help. Better still are the felt-bottom boots with little diamond-hard spikes.

I fish a lot, but my advice is, "Get a good guide." I had one in Andy, who in a very gentlemanly way pointed out my mistakes and helped me in every way. He's a good netman, a great fly adviser, and he got as big a kick when I got a fish as if he had taken it himself.

I find myself getting intolerant of those fishermen using hardware. There is something more sporting, more competitive, more difficult, more challenging about using a fly rod. . . .

I'm a very happy and a very lucky man now. Because of time spent fishing and the chance that fishing gives me to relax and think freely, now more than ever I see clearly just how blessed I really am. I served my country. I have a close family and wonderful wife for 52 and a half years, and, yes, I went to the Tree River and caught char.

Tight lines to all you fishermen!

Appendix D: “When You Bait the Hook with Your Heart, the Fish Always Bite”

Essay by John Burroughs

John Burroughs (1837–1921) was an American naturalist, author, and passionate lifelong fisherman from upstate New York. He hiked and fished through Alaska, Yellowstone, Yosemite, and the Grand Canyon; befriended Theodore Roosevelt, Henry Ford, and Thomas Edison; and inspired millions of Americans to love nature through his articles and books. In these excerpts from his essay “Speckled Trout,” from his book In the Catskills *(1910), he describes a fishing trip deep in the trout country of the Catskill Mountains.*

I have been a seeker of trout from my boyhood, and on all the expeditions in which this fish has been the ostensible purpose I have brought home more game than my creel showed. In fact, in my mature years I find I got more of nature into me, more of the woods, the wild, nearer to bird and beast, while threading my native streams for trout, than in almost any other way. It furnished a good excuse to go forth; it pitched one in the right key; it sent one through the fat and marrowy places of field and wood.

Then the fisherman has a harmless, preoccupied look; he is a kind of vagrant that nothing fears. He blends himself with the trees and the shadows. All his approaches are gentle and indirect. He times himself to the meandering, soliloquizing stream; its impulse bears him along. At the foot of the waterfall he sits sequestered and hidden in its volume of sound. The birds know he has no designs upon them, and the animals see that his mind is in the creek. His enthusiasm anneals him, and makes him pliable to the scenes and influences he moves among.

Then what acquaintance he makes with the stream! He addresses himself to it as a lover to his mistress; he wooes it and stays with it till he knows its most hidden secrets. It runs through his thoughts not less than through its banks there; he feels the fret and thrust of every bar and boulder. Where it deepens, his purpose deepens; where it is shallow, he is indifferent. He knows how to interpret its every glance and dimple; its beauty haunts him for days.

I am sure I run no risk of overpraising the charm and attractiveness of a well-fed trout stream, every drop of water in it as bright and pure as if the nymphs had brought it all the way from its source in crystal goblets, and as cool as if it had been hatched beneath a glacier. When the heated and soiled and jaded refugee from the city first sees one, he feels as if he would like to turn it into his bosom and let it flow through him a few hours, it suggests such healing freshness and newness. How his roily thoughts would run clear; how the sediment would go downstream! Could he ever have an impure or an unwholesome wish afterward? The next best thing he can do is to tramp along its banks and surrender himself to its influence. If he reads it intently enough, he will, in a measure, be taking it into his mind and heart, and experiencing its salutary ministrations.

Trout streams coursed through every valley my boyhood knew. I crossed them, and was often lured and detained by them, on my way to and from school. We bathed in them during the long summer noons, and felt for the trout under their banks. A holiday was a holiday indeed that brought permission to go fishing over on Rose's Brook, or up Hardscrabble, or in Meeker's Hollow; all-day trips, from morning till night, through meadows and pastures and beechen woods, wherever the shy, limpid stream led. What an appetite it developed! a hunger that was fierce and aboriginal, and that the wild strawberries we plucked as we crossed the hill teased rather than allayed. When but a few hours could be had, gained perhaps by doing some piece of work about the farm or garden in half the allotted time, the little creek that headed in the paternal domain was handy; when half a day was at one's disposal, there were the hemlocks, less than a mile distant, with their loitering, meditative, log-impeded stream and their dusky, fragrant depths.

Alert and wide-eyed, one picked his way along, startled now and then by the sudden bursting-up of the partridge, or by the whistling wings of the "dropping snip," pressing through the brush and the briers, or finding an easy passage over the trunk of a prostrate tree, carefully letting his hook down through some tangle into a still pool, or standing in some high, sombre avenue and watching his line float in and out amid the moss-covered boulders. In my first essayings I used to go to the edge of these hemlocks, seldom dipping into them beyond the first pool where the stream swept under the roots of two large trees. From this point I could look back into the sunlit fields where the cattle were grazing; beyond, all was gloom and mystery; the trout were black, and to my young imagination the silence and the shadows were blacker. But gradually I yielded to the fascination and penetrated the woods farther and farther on each expedition, till the heart of the mystery was fairly plucked out. During the second or third year of my piscatorial experience I went through them, and through the pasture and meadow beyond, and through another strip of hemlocks, to where the little stream joined the main creek of the valley.

In June, when my trout fever ran pretty high, and an auspicious day arrived, I would make a trip to a stream a couple of miles distant, that came down out of a comparatively new settlement. It was a rapid mountain brook presenting many difficult problems to the young angler, but a very enticing stream for all that, with its two saw-mill dams, its pretty cascades, its high, shelving rocks sheltering the mossy nests of the phoebe-bird, and its general wild and forbidding aspects.

But a meadow brook was always a favorite. The trout like meadows; doubtless their food is more abundant there, and, usually, the good hiding-places are more numerous. As soon as you strike a meadow the character of the creek changes: it goes slower and lies deeper; it tarries to enjoy the high, cool banks and to half hide beneath them; it loves the willows, or rather the willows love it and shelter it from the sun; its spring runs are kept cool by the overhanging grass, and the heavy turf that faces its open banks is not cut away by the sharp hoofs of the grazing cattle. Then there are the bobolinks and the starlings and the meadowlarks, always interested spectators of the angler; there are also the marsh marigolds, the buttercups, or the spotted lilies, and the good angler is always an interested spectator of them. In fact, the patches of meadow land that lie in the angler's course are like the happy experiences in his own life, or like the fine passages in the poem he is reading; the pasture oftener contains the shallow and monotonous places. In the small streams the cattle scare the fish, and soil their element and break down their retreats under the banks. Woodland alternates the best with meadow: the creek loves to burrow under the roots of a great tree, to scoop out a pool after leaping over the prostrate trunk of one, and to pause at the foot of a ledge of moss-covered rocks, with ice-cold water dripping down. How straight the current goes for the rock! Note its corrugated, muscular appearance; it strikes and glances off, but accumulates, deepens with well-defined eddies above and to one side; on the edge of these the trout lurk and spring upon their prey.

The angler learns that it is generally some obstacles or hindrances that makes a deep place in the creek, as in a brave life; and his ideal brook is one that lies in deep, well-defined banks, yet makes many a shift from right to left, meets with many rebuffs and adventures, hurled back upon itself by rocks, waylaid by snags and trees, tripped up by precipices, but sooner or later reposing under meadow banks, deepening and eddying beneath bridges, or prosperous and strong in some level stretch of cultivated land with great elms shading it here and there.

But I early learned that from almost any stream in a trout country the true angler could take trout, and that the great secret was this, that, whatever bait you used, worm, grasshopper, grub, or fly, there was one thing you must always put upon your hook, namely, your heart: when you bait your hook with your heart the fish always bite; they will jump clear from the water after it; they will dispute with each other over it; it is a morsel they love above everything else. With such bait I have seen the born angler (my grandfather was one) take a noble string of trout from the most unpromising waters, and on the most unpromising day. He used his hook so coyly and tenderly, he approached the fish with such address and insinuation, he divined the exact spot where they lay: if they were not eager, he humored them and seemed to steal by them; if they were playful and coquettish, he would suit his mood to theirs; if they were frank and sincere, he met them halfway; he was so patient and considerate, so entirely devoted to pleasing the critical trout, and so successful in his efforts,—surely his heart was upon his hook, and it was a tender, unctuous heart, too, as that of every angler is. How nicely he would measure the distance! How dexterously he would avoid an overhanging limb or bush and drop the line exactly in the right spot! Of course there was a pulse of feeling and sympathy to the extremity of that line.

If your heart is a stone, however, or an empty husk, there is no use to put it upon your hook; it will not tempt the fish; the bait must be quick and fresh. Indeed, a certain quality of youth is indispensable to the successful angler, a certain unworldliness and readiness to invest yourself in an enterprise that doesn't pay in the current coin. Not only is the angler, like the poet, born and not made, as Walton says, but there is a deal of the poet in him, and he is to be judged no more harshly; he is the victim of his genius: those wild streams, how they haunt him! He will play truant to dull care, and flee to them; their waters impart somewhat of their own perpetual youth to him. My grandfather when he was eighty years old would take down his pole as eagerly as any boy, and step off with wonderful elasticity toward the beloved streams; it used to try my young legs a good deal to follow him, specially on the return trip.

As it was then after two o'clock, and the distance was six or eight of these terrible hunters miles, we concluded to take a whole day to it, and wait till next morning. The Beaverkill flowed west, the Neversink south, and I had a mortal dread of getting entangled amid the mountains and valleys that lie in either angle.

Besides, I was glad of another and final opportunity to pay my respects to the finny tribes of the Neversink. At this point it was one of the finest trout streams I had ever beheld. It was so sparkling, its bed so free from sediment or impurities of any kind, that it had a new look, as if it had just come from the hand of its Creator. I tramped along its margin upward of a mile that afternoon, part of the time wading to my knees, and casting my hook, baited only with a trout's fin, to the opposite bank. Trout are real cannibals, and make no bones, and break none either, in lunching on each other. A friend of mine had several in his spring, when one day a large female trout gulped down one of her male friends, nearly one third her own size, and went around for two days with the tail of her liege lord protruding from her mouth! A fish's eye will do for bait, though the anal fin is better.

About four o'clock we reached the bank of a stream flowing west. Hail to the Beaverkill! and we pushed on along its banks. The trout were plenty, and rose quickly to the hook; but we held on our way, designing to go into camp about six o'clock. Many inviting places, first on one bank, then on the other, made us linger, till finally we reached a smooth, dry place overshadowed by balsam and hemlock, where the creek bent around a little flat, which was so entirely to our fancy that we unslung our knapsacks at once. While my companions were cutting wood and making other preparations for the night, it fell to my lot, as the most successful angler, to provide the trout for supper and breakfast. How shall I describe that wild, beautiful stream, with features so like those of all other mountain streams? And yet, as I saw it in the deep twilight of those woods on that June afternoon, with its steady, even flow, and its tranquil, many-voiced murmur, it made an impression upon my mind distinct and peculiar, fraught in an eminent degree with the charm of seclusion and remoteness.

The solitude was perfect, and I felt that strangeness and insignificance which the civilized man must always feel when opposing himself to such a vast scene of silence and wildness.

Appendix E: Iconic World Records by American Anglers

By Jack Vitek, World Records Coordinator, Internatonal Game Fish Association

MUSKELLUNGE

On the morning of July 24, 1949, more than sixty-five years ago, Cal Johnson and his son launched their boat in Lake Court Oreilles, located in their hometown of Hayward, Wisconsin. Not long after he started trolling a wooden Pike Oereno lure, as he had done for years, Johnson hooked up to what he immediately knew was a huge musky. Johnson, an experienced angler and outdoor writer, skillfully and carefully played his fish for an hour before the fish could be subdued. With the musky measuring more than five feet in length, Johnson knew he had something special. The musky was

George Perry, left. (Courtesy IGFA)

then taken to the nearby Moccasin Lodge, where it was officially weighed in at an enormous 30.62 kg (67 pounds, 8 ounces).

As is the case with most highly coveted awards, the All-Tackle record for musky has seen its share of controversy. Over the years, larger musky catches have been reported, such as Louie Spray's 69-pound,

David Hayes. (Courtesy IGFA)

11-ounce fish and Robert Malo's 70-pounder. However, Johnson's musky has retained the prestigious title as it was caught and documented in accordance with the IGFA's International Angling Rules—the internationally accepted rules of sportfishing.

LARGEMOUTH BASS

The All-Tackle record for largemouth bass is the most-sought-after game fish record in the world. It is the "holy grail" of fishing records. George Perry has held this prestigious title for nearly eighty-three years, since he pulled his massive 22-pound, 4-ounce fish from Montgomery Lake, Georgia, on June 2, 1932.

Perry, a twenty-year-old farmer at that time, decided to go fishing with longtime friend Jack Page. The two were taking turns with a single rod and reel, casting a Creek Chub Fintail Shiner from the wooden johnboat Perry had built. An interview from 1973 recorded Perry saying, "I thought I had hooked a log, but then the log started moving." After skillfully playing the fish out of a half-submerged treetop, Perry finally boated the fish, which was bigger than anything he or Page had ever seen. The two immediately beached the boat and headed for town. Later that day, the fish was officially weighed in at 22 pounds, 4 ounces and soon after became the new benchmark for record-chasing anglers around the world.

SMALLMOUTH BASS

In the world of freshwater fishing, especially in North America, there are few species more heavily targeted than the smallmouth bass. So it is no surprise that David Hayes's celebrated 5.41 kg (11 pounds, 15 ounces) smallmouth has seen its share of controversy over the years. Hayes caught his record fish on July 9, 1955, while trolling a lure in Dale Hollow Reservoir, Tennessee. Hayes held the All-Tackle title for forty-one years, despite swirling rumors denouncing his catch.

These rumors, coupled with an affidavit stating that the dock owner added lead weight to the catch (unbeknownst to Hayes), resulted in the temporary ousting of Hayes's record. However, nine years later, it was proven through multiple polygraph tests that the sworn affidavit that denounced the legitimacy of Hayes's smallmouth had been falsified. Thus, returning the All-Tackle title to Hayes. Despite the controversy surrounding Hayes's smallmouth, it has withstood the test of time—and quite a few polygraphs, too.

WHITE STURGEON

As Joey Pallotta III motored his 18-foot fiberglass boat toward San Pablo Bay, California, on the morning of July 9, 1983, little did he know that in just a few minutes, he would be fighting the biggest freshwater fish ever recorded on rod and reel. Just five minutes after casting a live grass shrimp into the water (remember that saying "elephants eat peanuts"?) Pallotta came tight on the massive 212.48 kg (468 pounds) white sturgeon that immediately surfaced, and began tail-walking just a few yards from Pallotta's boat. Stunned by the massive size of the fish, Pallotta radioed a nearby friend for assistance, as attempting to land such a fish on his boat was simply out of the question. Pallotta boarded

Joey Pallotta III. (Courtesy IGFA)

his friend's vessel, unassisted, and proceeded to fight the massive sturgeon for another five hours on 37 kg (80-pound) tackle, before it was finally subdued. There had never been an All-Tackle record for white sturgeon before Pallotta's, and there very well may never be another. Given the massive size of his fish and the strict regulations on the species, it is quite possible that Pallotta's fish will never be surpassed.

Sara Hayward. (Courtesy IGFA)

NILE PERCH

For many anglers, catching a perch is not something found on your angling "bucket list"—unless you're talking about Nile perch. Unlike their smaller relatives, Nile perch grow to incredible sizes and have earned a reputation as a vicious, no-nonsense adversary. These mighty fish are widespread throughout Africa, bringing anglers from around the world with hopes of landing one of these trophies. On December 20, 2000, California angler William Toth landed the heaviest Nile perch ever recorded on rod and reel—a 104.32 kg (230 pounds) fish that crushed the Rapala Fire Tiger lure he was trolling in Egypt's famous Lake Nasser. After a fight that lasted almost an hour, Toth and two local guides were able to weigh the fish in a sling and then released it alive. Toth's fish still remains the All-Tackle record despite the ever growing popularity of the Nile perch—a testament to the impressive nature of this catch.

WAHOO

The heaviest wahoo ever recorded by the IGFA—and which fish more resembles a World War II torpedo than a fish—was caught by Sara Hayward, a fifteen-year-old teenager from Texas, vacationing with her family in Cabo San Lucas, Mexico. Due to the size of the wahoo, and her young age, this one catch earned Hayward three different world records—the All-Tackle, the women's 37 kg (80 pounds) line class, and the Female Junior. Even more impressive is that the fish was caught on only 10.5 feet of 100-pound monofilament leader—no wire in sight! Hayward needed forty-five minutes to bring the massive wahoo to the boat, a 28-foot Californian captained by Gerry Martinez, named *Pez Espada II.* Hayward's wahoo eclipsed the previous All-Tackle record, which had stood since 1996, by nearly 30 pounds.

Alfred Glassell Jr. and record 1,560-pound black Marlin, August 4, 1953, Peru. (Courtesy IGFA)

Source Notes

INTRODUCTION

"Perhaps I should not have been a fisherman, he thought": Ernest Hemingway, *The Old Man and the Sea* (New York: Scribner, 1996), p. 43.

"All Americans believe that they are born fishermen": Warren G. French, *John Steinbeck's Nonfiction Revisited* (New York: Twayne, 1996), p. 85.

PROLOGUE

"Some go to church and think about fishing": Nick Lyons, *1,001 Pearls of Fishing Wisdom: Advice and Inspiration for Sea, Lake, and Stream* (New York: Skyhorse, 2013), p. 319.

"In our family, there was no clear line": Norman Maclean, *A River Runs Through it: And Other Stories* (New York: Pocket Books, 1992), p. 1.

"The trees, the shrubs, the flowers, the mosses": James Hart Hoadley, *Speckled Trout: Fishing Lines and Other Verse* (New York: T. E. Schulte, 1920), p. xiii.

"But how sweetly memories of the past come": Thaddeus Norris, *The American Angler's Book: Embracing the Natural History of Sporting Fish, and the Art of Taking Them* (Philadelphia: E. H. Butler, 1864), p. 28.

"No man ever steps in the same river twice": Duane Redford, *Fly Fisher's Playbook, 2nd Edition: A Systematic Approach to Nymphing* (Mechanicsburg, PA: Stackpole Books, 2015), p. iv.

"When you are on the river": Maxine Atherton, *The Fly Fisher and the River: A Memoir* (Skyhorse Publishing, 2016), p. 219.

"Come live with me, and be my love": John Donne, *Poems of John Donne,* vol. 1 (London: Lawrence & Bullen, 1896), p. 47.

"I will cast down my hook: The first fish which I bring up": *Catholic World,* vol. 75 (Paulist Fathers, 1902), p. 324.

"The first men that our Saviour dear": Izaak Walton, *The Compleat Angler; Or, Contemplative Man's Recreation; Being a Discourse on Rivers, Ponds, Fish and Fishing* (London: W. Orr, 1833), p. 94.

"Look at where Jesus went to pick people": Marc Folco, "Some hunting, fishing quotes to live by," *South Coast Today* (New Bedford, Massachusetts), July 18, 2015, accessed online.

"The traditional wilderness experience is an intense confrontation with God": Susan Power Bratton, *Christianity, Wilderness and Wildlife: The Original Desert Solitaire* (Scranton: University of Scranton Press, 1993), p. 242.

"One who believes that God made the world": Henry Ward Beecher, *Star Papers: Or, Experiences of Art and Nature* (New York: J. B. Ford, 1873), p. 234.

"Jesus is a fish that lives in the midst of waters": Richard Louv, *Fly-Fishing for Sharks: An American Journey* (New York: Simon & Schuster, 2002), p. 56.

"No life is so happy and so pleasant": Izaak Walton and Charles Cotton, *The Complete Angler, or, The*

Contemplative Man's Recreation: Being a Discourse of Rivers, Fish-ponds, Fish, and Fishing, vol. 2 (London: Nattali & Bond, 1860), p. 158.

"Time is but the stream I go a-fishing in": Henry David Thoreau, *Walden: A Fully Annotated Edition,* edited by Jeffrey S. Cramer (New Haven, CT: Yale University Press, 2004), p. 96.

"The best part of hunting and fishing was the thinking": Robert Ruark, *The Old Man and the Boy* (New York: Macmillan, 1993), p. 255.

"The angler who fishes carefully along the stream bank": Lefty Kreh, *Presenting the Fly* (New York: Lyons Press, 1998), p. 16.

"Every fisherman out there on the lake": Art Levy, "Roland Martin is a Florida Icon," *Florida Trend,* May 27, 2016.

1: THE FOUNDING FISHERMEN

"Only when the last tree has been felled": Nick Lyons, *1,001 Pearls of Fishing Wisdom: Advice and Inspiration for Sea, Lake, and Stream* (New York: Skyhorse, 2013), p. 295.

"At the falls of the rivers, where the water is shallow": *Smithsonian Miscellaneous Collections,* vol. 82 (Washington, DC: Smithsonian Institution, 1931), p. 140.

Mimbrenos: Stephen C. Jett and Peter B. Moyle, "The Exotic Origins of Fishes Depicted on Prehistoric Mimbres Pottery from New Mexico," *American Antiquity,* October 1986, pp. 688–720.

"To the native son, the shark was a horse to be bridled": *Arts and Crafts of Hawaii,* Bernice P. Bishop Museum Special Publication, no. 45 (Honolulu: Bishop Museum Press, 1957), p. 289.

"They are taken with a hook and line, but without any bait": Philip Alcemus Murray, *Fishing in the Carolinas* (Chapel Hill: University of North Carolina Press, 1941), p. 64.

"preserve codfish by hanging it in the frosty winter air": Mark Kurlansky, *Cod: A Biography of the Fish That Changed the World* (Toronto: Knopf Canada, 2011), p. 21.

"There they found self-sown fields of wheat": Anne Stine Ingstad and Helge Ingstad, *The Norse Discov-*

ery of America: The Historical Background and the Evidence of the Norse Settlement Discovered in Newfoundland (Oslo: Norwegian University Press, 1985), p. 528.

"We have won a fine and fruitful country": Ibid, p. 217.

"The sea there is swarming with fish which can be taken": Herbert John Wood, *Exploration and Discovery* (London and New York: Hutchinson's University Library, 1951), p. 72.

"To look into the depths of the sea is to behold the imagination": Victor Hugo, *The Works of Victor Hugo,* vol. 4 (New York: T. Y. Crowell, 1888), p. 187.

"The pleasant'st angling is to see the fish": William Shakespeare, "Much Ado About Nothing," in *The Works of Shakespeare,* vol. 1 (Oxford: Clarendon Press, 1771), p. 469.

"The water you touch in a river is the last": Vasiliĭ Pavlovich Zubov, *Leonardo da Vinci* (New York: Barnes & Noble, 1968), p. 221.

"The bass is one of the best fishes in the country"; "The halibut is not much unlike a plaice or turbot":

William Wood, *New England's Prospect,* edited by Alden T. Vaughan (Amherst: University of Massachusetts Press, 1993), p. 55.

"The Indians get many of them every day": William Wood, Alden T. Vaughan, *New England's Prospect* (University of Massachusetts Press, 1993), p. 56.

"no need to fish in deeper water": Callum Roberts, *The Unnatural History of the Sea* (Island Press, 2010) p. 37.

"At the south side of town there flows down": John Gorham Palfrey, *History of New England, Volume 1* (Little, Brown, 1859), p. 226.

"Under the laws of the British Empire, colonists were supposed to have sold their cod": Rebecca Fowler, "How Cod Created the World," *Daily Mail* (UK), July 22, 2000.

"By the mid-17th century, ships from Boston were delivering salt cod": Malabar Hornblower, "A Seafaring Past Preserved," *New York Times,* May 17, 1987, accessed online.

Philadelphia: A. J. Campbell, *Classic and Antique Fly-Fishing Tackle: A Guide for Collectors and Anglers* (Guilford, CT: Globe Pequot, 2002), pp. 6, 258.

2: THE FISHERMAN WHO CREATED AMERICA

George Washington's diary entries and quotes from letters in this chapter, unless otherwise noted, are from: National Archives, "Founders Online," http://founders.archives.gov. Some spelling and punctuation has been clarified.

"the larger part of the flesh diet of my people": Bill Mares, *Fishing with the Presidents* (New York: Stackpole Books, 1999), p. 5.

"which is black, is left for the blacks": Alexa Price, "Mount Vernon's Fisheries," George Washington's Mount Vernon website, http://www.mountvernon.org/digital-encyclopedia/article/mount-vernon-fisheries/.

"No pay! No clothes! No provisions!": "Revolutionary Battles Illustrated," *The Family Magazine, Or Monthly Abstract of General Knowledge,* vol. 3 (New York: J. S. Redfield, 1835), p. 234.

"no history now extant can furnish an instance": Paul Allen, John Neal, and Tobias Watkins, *A History of the American Revolution: Comprehending All the Principal Events Both in the Field and in the Cabinet* (Baltimore: F. Betts, 1822), p. 212.

Valley Forge research: John McPhee, *The Founding Fish* (New York: Macmillan, 2003), pp. 172–81.

"Then, dramatically, the famine completely ended": Ibid., p. 175.

Valley Forge research; an emergency alert was issued; "procure a large quantity recommend Shad Fish only"; "put up all the Fish you possibly can"; "you can render ten fold more service to your country"; "What success have you had in procuring shad?"; "this day I believe we will be nearly able to furnish": Joseph Lee Boyle, "The Valley Forge Fish Story," *Shad Journal* 4, no. 2 (1999), accessed online.

Sandy Hook: Paul F. Boller, *Presidential Diversions: Presidents at Play from George Washington to George W. Bush* (Boston: Houghton Mifflin Harcourt, 2007), pp. 4, 26.

Washington's love of fish dishes: Louisa May Skilton, "Washington—the Fish Dinner," *American Cookery,* February 1936, p. 377.

"On Monday, November 2nd, 1789, having lines": Charles Warren Brewster, William Henry Young Hackett, and Lawrence Shorey, *Rambles about Portsmouth: Sketches of Persons, Localities, and Incidents of Two Centuries: Principally from Tradition and Unpublished*

Documents, vol. 1 (Portsmouth, NH: C. W. Brewster & Son, 1859), p. 253.

3: THE GREATEST AMERICAN FISHING TRIP OF ALL

"There are always new places to go fishing": *Duluth News Tribune* website, January 5, 2014.

"Listen to the sound of the river and you will get a trout": T. O'Neill Lane, *Lane's English-Irish Dictionary* (London: D. Nutt, 1904), p. 241.

Quotations by Lewis and Clark in this chapter are from their journal entries cited in: University of Nebraska, Lincoln Libraries, Electronic Text Center, The Journals of the Lewis and Clark Expedition, http://lewisandclarkjournals.unl.edu. Some spelling and punctuation has been clarified.

"there's just something about casting for catfish": "20 Secrets to Help You Catch Fish All Summer Long," *Field & Stream* website, May 2013.

Catfish fishing tips: "20 Best Fishing Tips (Part 1)," http://www.twain2010.org/2015/12/23/10-best-fishing-tips/

"There are many kinds of natural beauty": Tim Eisele, "Book Dedicated to Trout," *The Capital Times* (Madison, Wisconsin), June 13, 1997.

"The salmon chief of the tribe would select a fisher": John Harrison, "First-Salmon Ceremony," Northwest Power and Conservation Council website, accessed July 23, 2016, http://www.nwcouncil.org/history/first salmonceremony.

"The salmon was put here by the Creator": "We Are All Salmon People," Columbia River Inter-Tribal Fish Commission website, accessed July 23, 2016, http://www.critfc.org/salmon-culture/we-are-all-salmon-people/.

4: THE FISH THAT WON THE LAST BATTLE OF THE CIVIL WAR

Joseph Wheelan, *Their Last Full Measure: The Final Days of the Civil War* (Boston: Da Capo Press, 2015), p. 176.

"They dimple, dapple, leap into the air": John McPhee, *The Founding Fish* (New York: Macmillan, 2003), p. 97.

"Of all the food fishes of America, shad may be regarded" *Boston Evening Transcript,* December 2, 1904, p. 13, quoting *The Philadelphia Record.*

"acoustic shadow," "food was abundant," "the affair was leisured and deliberate": Douglas Southall Freeman, *Lee's Lieutenants: Gettysburg to Appomattox* (New York: Charles Scribner's Sons, 1943), p. 668.

"Neither told any subordinate where he was going or why": Shelby Foote, *The Civil War, a Narrative: Red River to Appomattox* (New York: Vintage Books, 1986), p. 870.

"At that point, the Confederate forces": David Macaulay, "For Lee, Acoustics Were Sound Of Defeat," *Daily Press,* March 28, 2008, accessed July 23, 2016, http://articles.dailypress-com/2008-03-28/news/0803270357_1_confederate-acoustics-battle.

"Five Forks was the Army of the Potomac's first important battlefield victory"; "This has been the most momentous day of the war so far"; "Here is something material—something I can see": Joseph Wheelan, *Their Last Full Measure: The Final Days of the Civil War* (Boston: Da Capo Press, 2015), p. 176.

"Crook made the most of the terrain": John H. Monnett, "Mystery of the Bighorns: Did Fishing Trip Seal Custer's Fate?," *American Fly Fisher,* Fall 1993, p. 2.

"The streams around his base camp near the headwaters": Jim Merritt, "Custer Goes Hunting," *Field & Stream,* July 1999, p. 66.

"carried away by the desire to make a record"; "General Crook and the battalion commanders": "We tried them with all sorts of imported and manufactured flies"; "I gave him all the line he wanted": John Gregory Bourke, *On the Border with Crook* (New York: Charles Scribner's Sons, 1891), pp. 329–31.

Story of Frederick Benteen; "I saw him wade over his boot tops many times"; "Frederick Benteen remains, so far as is known": Richard Lessner, "How Meriwether Lewis's Cutthroat Trout Sealed Custer's Fate at the Little Bighorn," *American Fly Fisher,* Fall 2010, pp. 15–19.

5: AMERICA'S FLOATING INDUSTRY

"a speechlessly quick chaotic bundling of a man into eternity": Herman Melville, *Moby-Dick, or, The*

Whale (New York: Encyclopaedia Britannica, 1990), p. 17.

Nantucket sleigh rides; details of the sinking of the *Essex*: Gilbert King, "The True-Life Horror That Inspired Moby-Dick," Smithsonian.com, March 1, 2013, accessed July 23, 2016, http://smithsonianmag.com/history/the-true-life-horror-that-inspired-moby-dick-17576.

"The immense forehead of sperm whales is possibly the largest": Olga Panagiotopoulou, Panagiotis Spyridis, Hyab Mehari Abraha, David R. Carrier, and Todd C. Pataky, "Architecture of the sperm whale forehead facilitates ramming combat," *PeerJ*, April 5, 2016, accessed July 23, 2016, https://peerj.com/articles/1895/.

"The taking of one of a school, almost always ensures": Francis Allyn Olmsted, *Incidents of a Whaling Voyage* (New York: D. Appleton, 1841), p. 61.

Owen Chase memories and dialogue: Owen Chase, *Narrative of the Most Extraordinary and Distressing Shipwreck of the Whaleship Essex* (New York: Citadel Press, 1963), pp. 24–30.

"the most horrible and frightful convulsions I have ever witnessed": Philip Hoare, "When whales attack:

the horrific truth about Moby-Dick," *Telegraph,* December 26, 2015, accessed July 23, 2016, http://www.telegraph.co.uk/films/2016/04/14/when-whales-attack-the-horrific-truth-about-moby-dick/.

"humanity must shudder at the dreadful recital"; "separated limbs from his body"; "We knew not then to whose lot it would fall next": Adam Hodgson, *Letters from North America, Written During a Tour in the United States and Canada* (London: Hurst, Robinson, 1824), p. 381.

"My lad, my lad!"; "I like it as well as any other": "The Shipwrecked Mariners," *Sailor's Magazine and Naval Journal,* March 1832, p. 199.

"He was soon dispatched, and nothing of him left": Owen Chase, *Shipwreck of the Whaleship Essex* (New York: Lyons Press, 1999), p. 101.

"gastronomic incest": Brian Simpson, *Cannibalism and Common Law: A Victorian Yachting Tragedy* (London: A. & C. Black, 2003), p. 141.

"the most distressing narrative that ever came to my knowledge": Edouard A. Stackpole, *The Sea-hunters: the New England Whalemen During Two Centuries, 1635–1835* (Philadelphia: Lippincott, 1953), p. 330.

"What a commentator is this *Ann Alexander* whale": Laurie Robertson-Lorant, *Melville: A Biography* (Amherst: University of Massachusetts Press, 1998), p. 289.

"From the moment the Pilgrims landed": Eric Jay Dolin, *Leviathan, The History of Whaling in America* (W. W. Norton & Company, 2008), p. 11.

"Every day we saw whales playing hard by us": Moses Coit Tyler, *A History of American Literature During the Colonial Time, 1607–1676* (New York: G. P. Putnam's Sons, 1897), p. 160.

"lit the world and greased the gears": Dolin, *Leviathan,* inside cover copy.

"to produce a mighty book": Herman Melville, *Moby-Dick, Or, The Whale* (Penguin Group USA, 1955), p. 432.

"precisely as an orange is sometimes stripped": Herman Melville, *Moby-Dick* (New York: Harper & Brothers, 1851), p. 339.

"blood-stained decks, and the huge masses of flesh and blubber": Nathaniel Philbrick, *In the Heart of the Sea: The Tragedy of the Whaleship Essex* (New York: Penguin, 2001), p. 56.

"Processing a whale was nearly as dangerous as hunting one": "Whales and Hunting," New Bedford Whaling Museum website, https://www.whalingmuseum.org/learn/research-topics/overview-of-north-american-whaling/whales-hunting.

"They sailed the world's oceans and brought back tales": Eric Jay Dolin, *Leviathan: The History of Whaling in America* (New York: Norton, 2008), p. 12.

"A coloured man is only known and looked upon as a man": Ibid., p. 224.

"beef and bread one day, and bread and beef the next": Charles Lyman Newhall, *The Adventures of Jack: or, a Life on the Wave* (Printed by the author, 1859), p. 7.

"We have to work like horses and live like pigs": Briton Cooper Busch, *Whaling Will Never Do For Me: The American Whaleman in the Nineteenth Century* (Lexington: University Press of Kentucky, 2015), p. 14.

"black and slimy with filth, very small, and as hot as an oven": Foster Rhea Dulles, *Lowered Boats: A Chronicle of American Whaling* (New York: Harcourt, Brace, 1933), p. 84.

"the most filthy, indecent and distressed": Briton Cooper Busch, *Whaling Will Never Do For Me: The American Whaleman in the Nineteenth Century* (University Press of Kentucky, 2015), p. 13.

"Nowhere in all America will you find more patrician-like houses": Herman Melville, *Moby-Dick: Or, The Whale* (New York: Scribner, 1902), p. 28.

"Right now, it's a good story"; "This was the most impressive gray whale season": Clark Mason, "Whales Make Comeback off Sonoma Coast," Press-Democrat (Santa Rosa, CA), May 2, 2015, accessed online.

"A perilous life and sad as life can be": *Samuel Griswold Goodrich, Enterprise, Industry and Art of Man,* vol. 18 (Boston: Bradbury, Soden, 1845), p. 62.

"We have lingered in the chambers of the sea": T. S. Eliot, "The Love Song of J. Alfred Prufrock," in T. S. Eliot, *The Poems,* vol. 1 (London: Faber & Faber, 2015), p. 9.

"The history of the Gloucester fisheries has been written in tears": *The Fisheries of Gloucester from the First Catch by the English in 1623, to the Centennial Year, 1876* (Gloucester, MA: Procter Brothers, 1876), p. 71.

"When will the slaughter cease?": W. Jeffrey Bolster, *The Mortal Sea: Fishing the Atlantic in the Age of Sail* (Cambridge, MA: Harvard University Press, 2012), p. 229.

"were less likely to be suspected as runaways": *Frederick Douglass, Narrative of the Life of Frederick Douglass, an American Slave* (Globe Fearon, 1995), p. 66.

"The watermen, mostly black and some white": "American Shad and African American Watermen: Aspects of a Heritage Nearly Forgotten", presentation by Jim Cummins delivered at The Accokeek Foundation's African American Heritage Day, Sept 25, 2010, The Interstate Commission on the Potomac River Basin, http://www.potomacriver.org/wp-content/uploads/2014/12/afamwatermen.pdf

"Enclaves of free blacks": Harold Anderson, "Black Men, Blue Waters: African Americans on the Chesapeake," *The Star-Democrat* (Easton, Maryland), August 16, 1998.

"Many people don't realize it but hand tonging": paper by George Waters, quoted in blog post about S. Toranio Berry's documentary film *Black Captains of the Cheaspeake*: https://easternshorebrent.com/2016/04/15/black-captains-of-the-chesapeake-3/

Harold Anderson, "Black Men, Blue Waters: African Americans on the Chesapeake," Maryland Marine Notes Online, March-April 1998, http://ww2.mdsg.umd.edu/marinenotes/Mar-Apr98/index.php

"Whole communities grew around these industries": "On the Water: Commercial Fishers", Smithsonian Institution website, http://americanhistory.si.edu/onthewater/exhibition/3_3.html

"The waste during the seining season is enormous": Ibid., p. 118.

"is by far the worst weather I've ever seen": William Booth, "The Bering Sea's Dependable Haul: Tragedy," Washington Post, May 13, 2001, accessed online.

"Today's fishermen are descendants of the oldest continually operated business": Bolster, *The Mortal Sea,* p. 281.

"We have devastated cod by overwhelming their ecosystem": Callum Roberts, *The Unnatural History of the Sea* (Washington, DC: Island Press, 2010), p. 212.

"The twentieth century heralded an escalation in fishing": ibid., p. 364.

"The sea, once it casts its spell": Jacques-Ives Cousteau, *Life and Death in a Coral Sea* (A&W Visual Library, 1978), p. 13.

"The work of pulling creatures from the sea for a living": William B. McCloskey, *Their Father's Work: Casting Nets with the World's Fishermen* (Camden, ME: International Marine, 1998), p. 28.

"As long as there's one fish left": David Arnold, "Gloucester putting names to its fishing losses at sea," *Boston Globe*, March 19, 2000, accessed online.

6: THE GOLDEN AGE OF AMERICAN SPORTFISHING

"Never mind if the trout aren't biting": Tad Bartimus, Associated Press, "Eloquent Devotees to Fly Fishing in America," *Lawrence Journal-World,* June 12, 1988, p. 2C.

Background on Theodore Gordon: Gordon M. Wickstrom, "The Presence of Theodore Gordon," *American Fly Fisher,* Spring 2001, pp. 2–7.

"You know the Catskills, lad": James Fenimore Cooper, *The Cooper Gallery, or, Pages and Pictures from the Writings of James Fenimore Cooper* (New York: J. Miller, 1865), p. 123.

"The painters who were attracted to the Catskills": Alf Evers, *The Catskills: from Wilderness to Woodstock* (Garden City, NY: Doubleday, 1972), p. 397.

"The typical Catskill trout stream and its surroundings": James T. Yenckel, "The Catskills," *Washington Post,* February 12, 1984, accessed online.

"If I were a trout, I should ascend every stream": John Burroughs, *The Writings of John Burroughs,* vol. 1 (Boston: Houghton Mifflin, 1895), p. 152.

"a cranky old cuss": Justin Askins, *The Legendary Neversink: A Treasury of the Best Writing about One of America's Great Trout Rivers* (New York: Skyhorse, 2014), p. 158.

"We usually find that [fisher]men of the greatest experience": John Merwin, "The Legends of Fishing," *Field & Stream,* May 2005, p. 78.

"The great charm of fly-fishing is that we are always learning": Louis Bignami, *Wit & Wisdom of Fishing* (Lincolnwood, IL: Publications International, 1997), p. 37.

"Their sleek muscle-bound bodies cut through water": Callum Roberts, "Extinction Waiting to Happen," *Trust Magazine* (Pew Charitable Trust), Fall 2007 issue.

"What allows a tuna to generate": Carl Safina, *Song for the Blue Ocean: Encounters Along the World's Coasts and Beneath the Seas* (Macmillan, 2010), p. 55.

"caught the biggest fish of the crowd": Kenneth Schuyler Lynn, *Hemingway* (Cambridge, MA: Harvard University Press, 1995), p. 45.

"I would think of a trout stream I had fished when I was a boy": Paul Hendrickson, *Hemingway's Boat: Everything He Loved in Life, and Lost, 1934–1961* (New York: Vintage Books, 2012), p. 370.

"It is wild as the devil and the most wonderful trout fishing": Carlos Baker, ed., *Ernest Hemingway: Selected Letters, 1917–1961* (New York: Simon & Schuster, 2003), p. 26.

"All clear, no brush and the trout are in schools": H. Lea Lawrence, *Prowling Papa's Waters: A Hemingway Odyssey* (Atlanta: Longstreet Press, 1992), p. 30.

"good stuff for stories": Ibid., p. 23.

"When you started before daybreak": Peter Griffin, *Along with Youth: Hemingway, the Early Years* (New York: Oxford University Press, 1987), p. 30.

On the day of his first wedding . . . Hemingway was late: John O'Connor, "Before the Fame, There Was the Fishing," *New York Times,* October 4, 2015, p. TR10.

"leaped clear of the water and fell again": "Ernest Hemingway: Hall of Fame," International Game Fish Association website (undated).

"enter unabashed into the presence of the very elder gods": Ernest Hemingway, *Across the River and into the Trees* (New York: Charles Scribner's Sons, 1967), p. 17.

"Dad had a big nice trunk": Bartimus, "Eloquent Devotees to Fly Fishing in America," p. 2C.

"She is a really sturdy boat": Ernest Hemingway, *Marlin!* (Big Fish Books, 1992), p. 42.

"She is a really sturdy boat": Lawrence, *Prowling Papa's Waters,* p. 122.

"lovingly possessed her, rode her, fished her"; "*Pilar* represented this little encapsulated existence": "The

Old Man and the Boat: Hemingway's Quest for Peace," *PBS NewsHour,* October 6, 2011.

"could be a boor and a bully and an overly competitive jerk": "The Old Man and the Boat: Hemingway on the Pilar," National Public Radio, October 2, 2011.

"outriggers big enough to skip a ten-pound bait": Hemingway, *Marlin!,* p. 42.

Details, and quotes of Hemingway's fishing trip with priest: Stuart B. McIver, *Hemingway's Key West* (Sarasota, FL: Pineapple Press, 2002), pp. 31–33.

Hemingway's "pump and reel": Hilary Hemingway and Carlene Brennen, *Hemingway in Cuba* (New York: Rugged Land, 2003), p. 14.

"The secret is for the angler never to rest": Ernest Hemingway, *Hemingway on Fishing* (New York: Simon & Schuster, 2012), p. 144.

"A big-game fisherman might have counted himself blessed": Hendrickson, *Hemingway's Boat,* p. 194.

"Shot 27 in two weeks"; "Then just when we had him whipped and on the surface": Baker, ed., *Ernest Hemingway: Selected Letters, 1917–1961,* p. 416.

"through both legs with one hand while gaffing a shark": Ernest Hemingway, "On Being Shot Again: A Gulf Stream Letter," *Esquire,* June 1935, p. 25.

"truly the most wonderful damned thing I have ever been on": Michael S. Reynolds, *Hemingway: The 1930s* (New York: Norton, 1998), p. 93.

"the two most famous beards of their time": Hemingway and Brennen, *Hemingway in Cuba,* p. 123.

Emergency Fishing Kit details and quotes: "WWII Fishing Kits," igfa.org, International Game Fish Association website (undated).

Mary, Jim, and David Cabela quotes: courtesy of the Cabela family.

"You may find structure which at the moment": William K. Johnke, *The Behavior and Habits of Largemouth Bass* (Dorbil Publishing, 1995), quoted in http://www.umpquavalleybassmasters.com/bassbook.htm

"When you cast, stop halfway instead": "Bass Fishing Tips: How to Catch Bass: 10 Bass Fishing Secrets from Professional Anglers": http://www.discoverboating.com/resources/article.aspx?id=513

"I quickly grabbed my measuring board": Gary Laden, "One That Got Away," *Atlanta Constitution,* August 22, 1989.

"When that fish swam away": Don Ecker, "Big One That Got Away Stays With Anglers": *The Record* (New Jersey), August 12, 1987.

"I've guided on the New Fork River for twenty-five years": Tad Bartimus, Associated Press, "Fly Fishermen the Self-Anointed Elite of American Anglers," *Fort Scott Tribune,* June 1, 1988, p. 6B.

"Big game angling has a brief history": Tom Gifford, *Anglers and Muscleheads* (New York: Dutton, 1960), p. xi.

"as close to perfection as I've ever seen": "Personalities: Kevin VanDam," outdoor.com website (undated).

"There's no question in my mind that fishing is a science"; "It's the decision making": Bill Heavey, "Q&A: Bill Heavey Takes on Kevin VanDam," *Field & Stream* website (undated).

"My greatest strength can also be a weakness at times"; "The one big thing that I think would help a lot of people catch more fish": "20 Questions: Kevin VanDam," ESPN website, posted February 2, 2010.

"A lot of people talk about conservation": "New Fishing Encyclopedia Loaded With Facts and Figures," *Pittsburgh Post-Gazette,* February 20, 2000.

"All at once the water splashed everywhere": Bart Crabb, *The Quest for the World Record Bass* (Los Angeles: ProStar, 1997), p. 36.

"More than half the intense enjoyment of fly-fishing is derived": John J. Duffy, Alan Jon Fortney, and David Ernest Robinson, *Vermont, an Illustrated History* (Northridge, CA: Windsor, 1985), p. 245.

"They say you forget your troubles on a trout stream": Verna Craig Shelton, *Making Memories* (iUniverse, 2011), p. 154.

7: RISE OF THE GREAT WOMEN ANGLERS

"I followed mountain streams in search of those wary, fighty, cocky little brook trout": Carrie Foote Weeks, "My First Land-Locked Salmon," *Outing,* vol. 36, April–September 1990, p. 618.

"Simply shocking!": Mary Kelley Moore, "Casting a New Role," *Bangor Daily News,* March 22, 1995, p. C1.

"probably the most expensive and elaborate ever made": David McMurray, "Rivaling the Gentleman in the Gentle Art: The Authority of the Victorian Woman Angler," *Sport History Review,* November 2008, pp. 99–126.

"supple and lithe as a young tree": Tom Groening, "Casting for Commerce: 'Fly Rod' Crosby made Maine's Outdoors a Popular Destination," *Bangor Daily News,* October 21, 2000, accessed online.

"awed the wide-eyed New Yorkers": McMurray, "Rivaling the Gentleman in the Gentle Art," p. 107.

"Nobody in New York had seen anything just like me": Ibid., p. 108.

"She was the first champion of woman's rights in the hunting and fishing line": Julia A. Hunter and Earle G. Shettleworth, *Fly Rod Crosby: the Woman who Marketed Maine* (Gardiner, ME: Tilbury House, 2000), p. 54.

"large doses of the great outdoors": Moore, "Casting a New Role," p. C1.

"Here at a farmhouse I was to try the healing power of nature": Hunter and Shettleworth, *Fly Rod Crosby,* p. 4.

"in Cornelia's hands the fly rod became a magic wand": Lyla Foggia, "Reel Women: the World of Women Who Fish," *American Fly Fisher,* Spring 1996, p. 2.

"Her face is white, but her heart is the heart of a brave": Hunter and Shettleworth, *Fly Rod Crosby,* p. 40.

"She is as patient in whipping a stream in the Maine woods": McMurray, "Rivaling the Gentleman in the Gentle Art," p. 106.

"She is an expert and ardent sportswoman": "Women in Journalism," *New York Times,* May 25, 1893, p. 1.

"That's mighty good stuff": "'Fly Rod' Crosby: First Registered Guide in Maine," *Bangor Daily News,* March 26, 2008, p. B1.

"Fly Rod told running stories from column to column": Groening, "Casting for Commerce."

"If you want to give Maine and the Maine Central Railroad some advertising": Moore, "Casting a New Role," p. C1.

"You are a great young woman": McMurray, "Rivaling the Gentleman in the Gentle Art," p. 106.

"Taking advantage of her talent for showmanship": Foggia, "Reel Women," p. 3.

"Six to eight thousand attendance in the afternoon": Hunter and Shettleworth, *Fly Rod Crosby,* p. 21.

"will dress just as they do when on hunting and fishing trips": *New York Times,* March 14, 1897, p. 32.

"From the State of Maine there is to come": "From Stream and Forest," *New York Times,* March 15, 1896.

"she was toppling the shaky foundations of conventional arguments"; "keep Maine what it is today—the best hunting and fishing preserve in America": Moore, "Casting a New Role," p. C1.

"By the end of the nineteenth century": McMurray, "Rivaling the Gentleman in the Gentle Art," p. 101.

"prowess as a huntress and as a fisherwoman has been heralded": Hunter and Shettleworth, *Fly Rod Crosby,* p. 39.

"The two women had much in common as professional athletes": Ibid., p. 29.

"much sought after"; "inquiries of ladies who, like her, have a fondness for fishing": "From Stream and Forest," *New York Times,* March 15, 1896.

"How I wish more of the ladies would leave": Moore, "Casting a New Role," p. C1.

"Why should not a woman do her share of fishing": McMurray, "Rivaling the Gentleman in the Gentle Art," p. 111.

"There is a society of sixteen ladies": "The Oldest Club in America," *The Century Illustrated Monthly Magazine, May 1883 to October 1883,* pp. 545, 546.

"angling was the preferred outdoor pastime": David McMurray, " 'A Rod of Her Own:' Women and Angling in Victorian North America," masters degree thesis, University of Lethbridge, 1994, p. 139.

"Miss McBride's skill in fly tying has long been known": McMurray, "Rivaling the Gentleman in the Gentle Art," pp. 104–05.

"the story of the fly is, that one time, when this famous angler was fishing": Mary Orvis Marbury, *Favorite Flies and Their Histories* (Boston: Houghton Mifflin, 1892), p. 350.

"We are glad to know that there are some sportsman": McMurray, "Rivaling the Gentleman in the Gentle Art," p. 103.

"The first fish I caught on a fly rod was magic": "Diana Rudolph: Why Fly Fishing," American Museum of Fly

Fishing video, posted on YouTube, October 12, 2014, https://www.youtube.com/watch?v-63vsaz6COPQ.

"gallant fight of one hour and twenty-five minutes": "Mrs. Stagg's 20LB. Tarpon," *Forest and Stream,* May 14, 1891, p. 332.

"quickly became a nuisance around the clubhouse": George Reiger, *Profiles in Saltwater Angling* (Englewood Cliffs, NJ: Prentice-Hall, 1973), p. 129.

"At this point I thought I was crazy"; "I can't keep going on forever": Sara Houston Chisholm Farrington, *Women Can Fish* (New York: Coward-McCann, 1951), p. 33.

"Kip suggested we go chumming on our wedding trip": Ibid., p. 4.

"Perfection. The real record is to take the first one": Nelson Bryant, "Big Game Anglers Test Waters for Records," *New York Times,* January 5, 1981.

Kay Brodney in the Amazon and Brodney quotes: Clive Gammon, "Please, Don't Fall in the Water!," *Sports Illustrated,* May 18, 1981.

"She did it all"; "Kay fished in Argentina"; "We didn't carry any weapons": "The History Makers," American Museum of Fly Fishing website.

"She has done for casting what Stephen Hawking did for physics": Richard Louv, *Fly-Fishing for Sharks: An American Journey* (New York: Simon & Schuster, 2002), p. 246.

"I loved fly casting because of the grace and beauty of it": Danielle Ibister, *The Fly Fishing Anthology* (Stillwater, MN: Voyageur Press, 2004), p. 147.

"As is true of many of the women of my generation": Joan Wulff, *Joan Wulff's Fly Fishing: Expert Advice from a Woman's Perspective* (New York: Stackpole Books, 1991), p. vi.

8: FISHERMEN IN THE MODERN WHITE HOUSE

"Fish are constantly doing the most mysterious and startling things": Grover Cleveland, *Fishing and Shooting Sketches* (New York: Outing, 1906), p. 28.

FDR's visit to Antoine's restaurant, savored *Pompano en Papillote:* Murphy Givens, "Roosevelt Went Tarpon Fishing off Port Aransas," *Corpus Christi Caller Times,* March 26, 2014, accessed online.

"rainwater soup, rubber squab": Carlos Baker, ed., Ernest Hemingway*: Selected Letters 1917–1961* (New York: Simon & Schuster, 2003), p. 460.

"stood over the cooks, making sure": Laura Shapiro, "The First Kitchen," *The New Yorker,* November 22, 2010.

"one of the most highly prized gulf coast fish": David Hagedorn, "Pompano en Papillotte With Crawfish Sauce," *Washington Post*, October 27, 2010, accessed online.

"We are having a most delightful cruise": Elliott Roosevelt, *F.D.R.: His Personal Letters: 1928–1945* (New York: Duell, Sloan & Pearce, 1950), p. 512.

"We hugged azure skies, golden sands, turquoise depths"; "Roosevelt was a first-class fisherman, who usually was successful": Robert F. Cross, *Sailor in the White House: The Seafaring Life of FDR* (Annapolis, MD: Naval Institute Press, 2014), p. 96.

"He could fish from the back of pleasure boats": Bill Mares, *Fishing with the Presidents* (New York: Stackpole Books, 1999), p. 198.

"The hook had torn a large hole in the tarpon's mouth"; "I am going to lose your first tarpon"; "When we got there we were faced with the problem"; "What in the world?": Barney Farley, George S. Hawn, and Larry McEachron, *Fishing Yesterday's Gulf Coast* (College Station: Texas A&M University Press, 2008), pp. 8–10.

"On Sunday the President wanted to fish in a stream": Winston Churchill, *The Second World War,* vol. 4, *World War, 1939–1945* (London: Cassell, 1951), p. 713.

"The two men sat side by side on portable canvas chairs": William McKinley Rigdon, *White House Sailor* (Garden City, NY: Doubleday, 1962), p. 219.

"over a distance of three and one half miles"; "Thanks old fellow, you put up a good fight": Mares, *Fishing with the Presidents,* p. 176.

"In rippling mountain brooks and lazy southern ponds": Lawrence Knutson, "The Ways of Fish and Presidents," Associated Press, *Ludington Daily News,* April 20, 1999, p. 4.

"We caught only a few small fish and had the pleasure of rowing": John Quincy Adams, *Diary of John*

Quincy Adams, March 1786–December 1788, vol. 2 (Cambridge, MA: Belknap Press of Harvard University Press, 1981), p. 216.

"I had been fishing one day and caught a little fish": John Nicholay and John Hay, "Abraham Lincoln: a History," *Century,* November 1886, p. 16.

"There is nothing I loved more than fishing for salmon": "Sportsman Channel Honors Great Eight 'Sportsman' U.S. Presidents," *Outdoor Wire,* February 11, 2010.

"No man can pitch a tent more quickly": George Dawson, *Angling Talks: Being the Winter Talks on Summer Pastimes* (New York: Forest and Stream Publishing Company, 1883), p. 41.

"I may be president of the United States": Justus D. Doenecke, *The Presidencies of James A. Garfield & Chester A. Arthur* (Lawrence: Regents Press of Kansas, 1981), p. 79.

"Grover Cleveland will fish when it shines and fish when it rains": Mares, *Fishing with the Presidents,* p. 65.

"There can be no doubt that certain men are endowed": Grover Cleveland, *Fishing and Shooting Sketches* (NY: Outing Publishing Co., 1906), p. 7.

"In many cases the encounter with a large fish": Ibid., p. 34.

"What sense is there in the charge of laziness": Ibid., p. 26.

"It must, of course, be admitted that large stories of fishing adventure"; "Beyond these presumptions, we have the deliberate and simple story": Grover Cleveland, "A Defense of Fishermen," *Saturday Evening Post,* October 19, 1901, pp. 3, 4.

"The real worth and genuineness of the human heart": Grover Cleveland, "The Mission of Fishing and Fishermen," *Saturday Evening Post,* December 5, 1903, p. 1.

Benjamin Harrison spit on his worms: Edmund Morris, *The Rise of Theodore Roosevelt* (New York: Random House, 2010), p. 406.

Details and quotes of Theodore Roosevelt's manta hunting expedition: Theodore Roosevelt, "Harpooning Devilfish," *Scribner's,* September 1917, pp. 293–305.

"detested fishing": Paul Schullery, "Theodore Roosevelt as an Angler," *American Fly Fisher,* Summer 1982, p. 21.

1871 trip to upstate New York, father reading *The Last of the Mohicans:* Theodore Roosevelt, *Theodore Roosevelt's America: Selections from the Writings of the Oyster Bay Naturalist* (New York: Anchor Books, 1955), p. xxvi.

"Sometimes we vary our diet with fish": Theodore Roosevelt, "The Home Ranch," *Century Illustrated Magazine,* March 1888, p. 666.

"Around this camp there was very little game": Theodore Roosevelt, *The Works of Theodore Roosevelt: Hunting Trips of a Ranchman* (New York: Charles Scribner's Sons, 1906), p. 128.

"The President never fishes unless put to it for meat": John Burroughs, "Camping with President Theodore Roosevelt," *Atlantic Monthly,* vol. 97 (1906), p. 595.

"Even if Roosevelt had never fished in his life": Mares, *Fishing with the Presidents,* p. 33.

"Don't think I'll go fishing"; "Fishing is for small boys": Lyle C. Wilson, United Press, "Hunting Trip for Ike Recalls Marksmanship of Col. Starling," *Times-News* (Hendersonville, NC), February 17, 1956.

"Mama! Mama!"; "Look what I've caught!": Edmund Starling, "Presidents off guard," *Life,* vol. 20, no. 7, February 18, 1946, p. 112.

"Shall I use the whole worm?": Mares, *Fishing with the Presidents,* p. 197.

"Damn!"; "Guess I'm a real fisherman now": Ibid., p. 157.

"I haven't caught them all yet, but I've intimidated them": Clifton Fadiman and Andre Bernard, *Bartlett's Book of Anecdotes* (Boston: Little, Brown, 2000), p. 140.

"Where is the salmon?"; "I sent it to the White House kitchen"; "The directors of the fishing club, the fish and I posed": Herbert Hoover, *Fishing for Fun—and to Wash Your Soul* (New York: Random House, 1963), p. 81.

"I think I have discovered the reason": Ibid., p. 76.

"It is the chance to wash one's soul with pure air": Ibid., p. 11.

"Where is the salmon?": Herbert Hoover, "Salmon and the Cotton," *Pittsburgh Post-Gazette,* August 8, 1963.

"Yeah, we heard back then that Hoover liked to fish": Howell Raines, "Fishing with Presidents," *New York Times Magazine*, September 5, 1993.

"He grabbed those fish and held them up real fast": Mares, *Fishing with the Presidents*, p. 44.

"The gatherings were always a small group of men": Dwight D. Eisenhower, *The White House Years: Mandate for Change, 1953–1956*, vol. 1 (Garden City, NY: Doubleday, 1963), p. 268.

"After hooking a limb the first three times"; "I could see that he was disappointed": Richard M. Nixon, *In the Arena: A Memoir of Victory, Defeat, and Renewal* (New York: Pocket Books, 1991), pp. 186–87.

Eisenhower's fishing in and around Abilene: Catherine Schmitt, *The President's Salmon: Restoring the King of Fish and its Home Waters* (Camden, ME: Down East Books, 2015), p. 89.

"I don't use worms": *AP Online*, April 19, 1999.

"There is a small stream on which we catch ten and twelve inchers"; Hoover-Eisenhower fishing story: Hal Elliott Wert, *Hoover, the Fishing President: Portrait of the Private Man and His Life Outdoors* (New York: Stackpole Books, 2005), pp. 314–16.

"Little Boy Blue": Richard Reeves, *President Kennedy: Profile of Power* (New York: Simon & Schuster, 1994), p. 22.

"Nixon wouldn't do anything but watch": Ronald Kessler, *In the President's Secret Service: Behind the Scenes with Agents in the Line of Fire and the Presidents They Protect* (New York: Crown, 2009), p. 31.

"As a youngster I reluctantly fished": Mares, *Fishing with the Presidents,* p. 17.

"I've lost the fish, Daddy": Jimmy Carter, *An Hour Before Daylight: Memories of a Rural Boyhood* (New York: Simon & Schuster, 2001), p. 107.

"I had a fishing pole in my hands as early as I can remember": Ibid., p. 97.

"one of the most gratifying developments": Jimmy Carter, *An Outdoor Journal: Adventures and Reflections* (Little Rock: University of Arkansas Press, 1994), p. 15.

"It levels people out"; "I really enjoy fishing with a plastic worm for bass": Raines, "Fishing With Presidents."

"I took along with me the briefing books": Tad Bartimus, "Lure of the Fly," *The Record* (New Jersey), August 28, 1988.

Carter took fifty-two fishing trips: "President Carter's Trips as President," Jimmy Carter Presidential Library, http://www.jimmycarterlibrary.gov/documents/jec/trips.phtml.

Carter's fly-fishing seminar day at Camp David: *The Daily Diary of President Jimmy Carter,* November 22, 1980, Jimmy Carter Presidential Library, http://www.jimmycarterlibrary.gov/documents/diary/1980/d112280t.pdf.

"I started fishing at age five"; "catching mackerel on a lead jig"; "trolling with that old green cotton line": Mares, *Fishing with the Presidents,* p. 19.

"the wily bonefish": Maureen Dowd, "An Angler in Paradise," *New York Times,* January 16, 1989.

"He was a troller, a spinner, a bait-caster": Catherine Schmitt, "Politics, Preservation, and Salmon Fishing," *Boston Globe,* May 17, 2015.

"You take them in, weigh them": Associated Press, January 15, 1989.

"I love the water. I can concentrate"; compared fishing to "heaven": George H.W. Bush, *All the Best, George Bush: My Life in Letters and Other Writings* (New York: Simon & Schuster, 2013), p. 617.

"I am hard-pressed to think of an outdoor activity": Recreational Boating & Fishing Foundation press release, PR Newswire, April 18, 2016.

"You're in another world": Polly Whittell, "On Board With George Bush," *Motor Boating & Sailing,* March 1988.

"Way above the tree line": George Bush, "The Thrill of Northern Fishing," column for *Deh Cho Drum* (Northwest Territories, Canada), September 4, 1997.

"Just let me know what you need"; "When Putin arrived on July 1, 2007": George W. Bush, *41: A Portrait of My Father* (New York: Crown/Archetype, 2014), p. 201.

"The weather on the day of fishing with President Obama": Pat Hill, "Fly Fishing With President Obama," *Montana Pioneer,* September 2009.

"You may be president": Carter, *An Outdoor Journal,* p. 15.

9: THE POWER OF FISHING

"Man's life is but vain; for 'tis subject to pain": Izaak Walton, Charles Cotton, The Complete Angler (Nattali and Bond, 1860), p. 265.

"In the night I dreamed of trout-fishing": Henry David Thoreau, "Ktaadn, and the Maine Woods," Sartain's Union Magazine of Literature and Art, Volumes 2–3, 1848, p. 238.

"Other than being in Vietnam and seeing people": "Wounded Warriors Go Fishing for Recovery," "CNN Heroes," CNN, April 11, 2009.

"I should take a couple of these guys fishing": Mark Yost, "Project Healing Waters," Wall Street Journal, August 15, 2012.

"The goal is simple": Angus Phillips, "Injured Veterans Find Therapy in Fly-Fishing," Washington Post, May 11, 2008.

"When we started this, I thought it would just be great": Bill May, "Veterans Healing Veterans," Carroll County Times (Maryland), May 30, 2016.

"It's a wonderful thing to behold": "Positive, Lasting Changes," Project Healing Waters website, http://www.projecthealingwaters.org/NewsMedia/Announcements/tabid/147/post/positive-lasting-changes-a-spouse-s-testimonial-for-project-healing-waters-fly-fishing/Default.aspx

"When I came back from Korea": Bill May, "Project Healing Waters, a Salute to Heroes," Carroll County Times (Maryland), May 30, 2016.

"When you're fly fishing or fly tying": Luanne Rife, "Salem Branch of Project Healing Waters Teaches Disabled Veterans Fly Fishing," The Roanoke Times (Virginia), April 19, 2015.

"All you could hear was the sound of the river"; "This program does a lot more": Charles Sowell, "Healing Waters Fishing Program Serves Military Vets," GreenvilleOnline (South Carolina), September 23, 2014.

"It helps me out, and it helps someone else"; "You've got to be very imaginative": Norman Moody, "Helping Disabled Veterans Heal Through Fly Fishing," Sun Sentinel (South Florida), March 8, 2015.

"This program is the best thing that ever happened": Wall Street Journal, August 15, 2012.

"It was almost hypnotic": Zach Benoit, "Casting Community: Fly Fishing Retreat Helps Breast Cancer Survivors Heal," Billings Gazette (Montana), February 2, 2016.

"It focuses you in on the moment": Amy Donaldson, "Women find Casting for Recovery Heals More than

Scars," Deseret News (Salt Lake City, Utah), September 11, 2011.

"magical" experience; "but the motion of casting a rod": Lisa Esposito, "Cancer Retreats: Finding a Fresh Perspective," U.S. News & World Report, June 3, 2015.

"It was such a Zen-like experience"; "I hadn't fished since sixth grade": Erin Andersen, "Breast Cancer Survivors Find Healing, Camaraderie in Snake River," Lincoln Journal Star (Lincoln, Nebraska), July 15, 2013.

"God grant that I may fish, until my dying day": Bassmaster News obituary for Homer Circle, June 26, 2012, https://www.bassmaster.com/news/homer-circle-dies-97

About the Authors

Willie Robertson is the CEO and resident prankster of Duck Commander, Buck Commander, Fin Commander, and Strut Commander, and author of the #1 *New York Times* bestseller *The Duck Commander Family.*

Armed with a business degree, he took the company from a living room operation, down by the river, to a premier destination for all things outdoors. He honed his skills as a salesman by selling his freshly caught fish at the market with his mom as a young boy. Even then, he always worked to negotiate the best price.

Willie is cast member and executive producer of both *Duck Dynasty* and *Buck Commander,* and exudes a passion for inspiring future hunters by showing the outdoor lifestyle and hunting experience in a fun and entertaining way. His unique perspective helps the two

TV shows, DVDs, and other products stand out in the crowd. While the Robertsons' story is a great example of entrepreneurship and following your passion, it's also about love, faith, and putting family first.

Willie loves being outdoors with his family and friends and is happiest at home in West Monroe, Louisiana, with his wife, Korie, and their children: John Luke, Sadie, Will, Bella, Rebecca, and Rowdy. When he is not working, you will catch him fishing, hunting, or on his tractor bush-hogging a field—earphones blaring, thinking of the next idea that will keep his companies at the top of the heap.

William Doyle is author of *PT 109: An American Epic of War, Survival, and the Destiny of John F. Kennedy*, and coauthor of *Navy SEALs: Their Untold Story* and co-producer of the companion PBS documentary film. He also coauthored *American Gun: A History of the U.S. in Ten Firearms*, with Chris Kyle.

Willie Robertson and William Doyle are coauthors of the *New York Times* bestseller *American Hunter: How Legendary Hunters Shaped America.*